Hippolyte Dussauce

Treatise on the Coloring Matters Derived from Coal Tar

their practical application in dyeing cotton, wool, and silk

Hippolyte Dussauce

Treatise on the Coloring Matters Derived from Coal Tar
their practical application in dyeing cotton, wool, and silk

ISBN/EAN: 9783337072018

Printed in Europe, USA, Canada, Australia, Japan

Cover: Foto ©berggeist007 / pixelio.de

More available books at **www.hansebooks.com**

TREATISE

ON THE

COLORING MATTERS DERIVED FROM COAL TAR;

THEIR

Practical Application in Dyeing Cotton, Wool, and Silk.

THE

PRINCIPLES OF THE ART OF DYEING AND THE DISTILLATION OF COAL TAR.

WITH A DESCRIPTION OF THE

MOST IMPORTANT NEW DYES NOW IN USE.

BY

Professor H. DUSSAUCE, Chemist,

Lately of the Laboratories of the French Government, viz., the Mining, Botanical Garden, the Imperial Manufacture of the Gobelins, the Conservatoire Impériale of Arts and Manufactures, Professor of Industrial Chemistry to the Polytechnic Institute, Paris.

PHILADELPHIA:
HENRY CAREY BAIRD,
INDUSTRIAL PUBLISHER
406 Walnut Street.

1863.

Entered according to Act of Congress, in the year 1863, by
HENRY CAREY BAIRD,
in the Clerk's Office of the District Court of the United States in and for the Eastern District of Pennsylvania.

PHILADELPHIA:
COLLINS, PRINTER.

PREFACE.

The object of this work is not to present to the public a treatise on the art of dyeing, but simply to furnish a full and clear description of those colors concerning which so much is said and so little known.

The greater part of the coloring matters employed by dyers, belongs to the vegetable, a few to the mineral, and fewer still to the animal kingdoms. Within a few years past a great variety of colors, among which are crimson, red, violet, blue, green, scarlet and yellow, have been obtained from a single substance—*coal tar*—and the shades produced on silk and wool by these colors are unrivalled in beauty.

As yet no distinct treatise on this subject has been published; all the information we have is found scattered here and there through many scientific and industrial publications, and thus rendered almost inaccessible to the practitioner. Our object has been to collect these scattered items, and, in connection with our knowledge of the subject, prepare a practical work for the dyer and calico printer, authors having devoted themselves more to the theory than the practice. The manipulations described in the different journals are difficult, and the great number of formulæ used, render the explanation unintelligible to any one not acquainted with chemical theories. We have endeavored to so simplify the recipes and minutely describe the manipulations as to enable any one, though not a chemist, to manufacture these colors. The principal works which we have consulted are: *Les Comptes Rendus, Annales de Chimie et de Physique, Bulletin de la Société d'encouragement, Moniteur industriel, The Chemical News, London Journal of Pharmacy, American Druggist's Circular*, etc. etc.

The book is divided into several chapters. The first is devoted to the general notions of the art of dyeing; several treat of the fabrication of colors of *coal tar* and their applications; and we terminate by the processes to manufacture different new colors, and the theory of the fixation of colors and mordants.

This work, the only one of the kind thus far published, we trust is destined to render great services to the dyer by removing the uncertainties now attending this new branch of industry, and enabling the dyer himself to manufacture those colors which he is now obliged to purchase at a very high price.

The approbation of the profession will be our most satisfactory reward.

THE AUTHOR.

New Lebanon, N. Y.,
April, 1863.

CONTENTS.

CHAPTER I.
                                          PAGE
Historical notice of the art of dyeing 25

CHAPTER II.
Chemical principles of the art of dyeing . . . 33

CHAPTER III.
Preliminary preparation of stuffs 39

CHAPTER IV.
Mordants 43

CHAPTER V.
Dyeing 47

CHAPTER VI.
On the coloring matters produced by coal tar . . 49

CHAPTER VII.
Distillation of coal tar 52

CHAPTER VIII.
History of aniline—Properties of aniline—Preparation of aniline directly from coal tar 60

CHAPTER IX.

Artificial preparation of aniline—Preparation of benzole—Properties of benzole—Preparation of nitro-benzole—Transformation of nitro-benzole into aniline, by means of sulphide of ammonium; by nascent hydrogen; by acetate of iron; and by arsenite of potash—Properties of the bi-nitro-benzole 68

CHAPTER X.

Aniline purple—Violine—Roseine—Emeraldine—Bleu de Paris 81

CHAPTER XI.

Futschine, or magenta 92

CHAPTER XII.

Coloring matters obtained by other bases from coal tar—Nitroso-phenyline—Di-nitro-aniline—Nitro-phenyline—Picric acid—Rosolic acid—Quinoline . . 98

CHAPTER XIII.

Naphthaline colors—Chloroxynaphthalic and perchloroxynaphthalic acids—Carminaphtha—Ninaphthalamine—Nitrosonaphthaline—Naphthamein—Tar red—Azuline 104

CHAPTER XIV.

Application of coal tar colors to the art of dyeing and calico printing 112

CHAPTER XV.

Action of light on coloring matters from coal tar . . 120

CHAPTER XVI.

Latest improvements in the art of dyeing—Chrysammic acid—Molybdic and picric acids—Extract of madder . . 125

CHAPTER XVII.

Theory of the fixation of coloring matters in dyeing and printing 133

CHAPTER XVIII.

Principles of the action of the most important mordants . 144

CHAPTER XIX.

Aluminous mordants 148

CHAPTER XX.

Ferruginous mordants 159

CHAPTER XXI.

Stanniferous mordants 170

CHAPTER XXII.

Artificial alizarin 175

CHAPTER XXIII.

Metallic hyposulphites as mordants—Dyer's soap—Preparation of indigo for dyeing and printing—Relative value of indigo—Chinese green—Murexide . . 179

COLORING MATTERS FROM COAL TAR.

CHAPTER I.

HISTORICAL NOTICE OF THE ART OF DYEING.

THE art of Dyeing has been successfully practised in the East Indies, Persia, Egypt, and Syria, from time immemorial. In the Pentateuch, frequent mention is made of linen cloths dyed blue, purple, and scarlet; and of ram skins dyed red; the works of the Tabernacle, and the vestments of the High Priest were commanded to be of purple.

The Tyrians were probably the only people of antiquity who made dyeing their chief occupation, and the staple of their commerce. The opulence of Tyre seems to have proceeded, in a great measure, from the sale of its rich and durable purple. It is unanimously asserted by all writers, that a Tyrian was the inventor of the purple dye, about 1500 years before the birth of CHRIST, and that the King of Phœnicia was so captivated with the color, that he made purple one of his principal ornaments, and that, for many centuries after, Tyrian purple became a badge of royalty. So highly prized was this color, that in the time of Augustus, a pound of wool dyed with it, cost at Rome, a sum nearly equal to thirty pounds sterling. The Tyrian purple is now generally believed to have been derived from two different kinds of shell-fish, described by Pliny under the names *purpura* and *buccinum*, and was extracted from a small vessel or sac in their throats to the amount of one drop from each animal; but an inferior substance was obtained by crushing the whole substance of the *buccinum*. At first it is a colorless liquid, but by exposure to air and light it assumes successively a citron-yellow, green, azure, red, and, in the course of forty-eight hours, a brilliant purple hue. If the liquid be evaporated to dryness soon after being collected, the residue does not become tinged in this manner. These circumstances correspond with the minute description of the manner of catching the purple-dye fish given in the work of an eye-witness, Eudocia Macrembolitissa, daughter of the Emperor Constantine the Eighth, who lived in the eleventh century. The color is remarkable for its durability. Plutarch observes, in his life of Alexander, that, at the taking of Susa, the Greeks found, in the royal treasury of Darius, a quantity of purple cloth, of the value of five thousand talents, which still retained its beauty, though it had lain there one hundred and ninety years. This color resists the action even of alkalies, and most acids.

Pliny states that the Tyrians gave the first ground of their purple dye by the unprepared liquor of the *purpura*, and then improved or heightened it by the liquor of the *buccinum*. In this manner they prepared their double-dyed purple—*purpura dibapha*—which was so called, either because it was immersed in two different liquors, or because it was first dyed in the wool and then in the yarn.

In ancient Greece it does not appear that the art of dyeing was much cultivated. In Rome it received more attention; but very little is now known of the processes followed by the Romans, such arts being held, by them, in low estimation. The principal ingredients used by these people were the following: Of vegetal matters, alkanet, archil, broom, madder, nutgalls, woad, and the seeds of the pomegranate, and of an Egyptian acacia; and of mineral productions, sulphate of iron, sulphate of copper, and a native alum mixed with the former.

The progress of dyeing, as of all other arts, was completely stopped in Europe, for a considerable time, by the invasion of the Northern barbarians in the fifth century. In the East the art still continued to flourish, but it did not revive in Europe until towards the end of the twelfth or the beginning of the thirteenth century. One of the places chiefly celebrated for this art was Florence, where, it is said, there were no less than two hundred establishments at work in the early part of the fourteenth century. A Florentine dyer, having ascertained in the Levant a method of extracting a coloring principle from the lichens which furnish archil, introduced this on his return, and acquired by its sale an immense fortune.

The discovery of America tended greatly to the advancement of this art, as the dyers were supplied thence with several valuable coloring materials previously unknown; amongst which are logwood, quercitron, Brazil-wood, cochineal, and annotto. About the year 1650, also, a great improvement in dyeing took place, which consisted in the introduction of a salt of tin as an occasional substitute for alum. With cochineal, the former was found to afford a color far surpassing in brilliancy any of the ancient dyes. To Cornelius Drebbel the merit of this application is attributed. His son-in-law established an extensive dye-house at Bow, near London, about the year 1563.

For several centuries the Italians, and particularly the Venetians, prosecuted the art of dyeing to a large extent, and long held a complete monopoly of the art, and procured large sums by it from other nations. In the year 1548, one John Ventura Rosetti published a book, termed *Plictho's Art of Dyeing*, in which he teaches how to give to cloth, linen, cotton and silk, real and beautiful, as well as false and common dyes, which is, perhaps, the first book that ever appeared upon the subject, and laid the first foundation for the improvement of this art which afterwards took place; it having excited the French, English, and Germans to apply in earnest, in their different countries, to improving so useful and extensive a branch of manufacture.

After this period the art was extensively carried on by the Flemings, and many of them emigrating to Germany, France, and England, established themselves as dyers, and thus gave great impetus to its advancement. In 1667, a Fleming named Brauer came to England with his whole family, and brought the dyeing of woollen there to that degree of perfection at which it has been ever since maintained. Shortly after this several works were published upon the art, which did much to improve it and make it more cultivated.

Logwood and indigo began to be employed as dyes in Europe about the middle of the sixteenth century, but not without considerable opposition from the cultivators of the native woad; the former were prohibited in England by Queen Elizabeth, under a very heavy penalty, and all found in the country was ordered to be destroyed: their use was not permitted till the reign of Charles the Second.

Indigo, the innoxious and beautiful product of

an interesting tribe of tropical plants, which is adapted to form the most useful and substantial of all dyes, was actually denounced as a dangerous drug—*food for the devil*, it was called—and forbidden by Parliament, in the reign of Elizabeth, to be used. An act was passed, authorizing searchers to burn both it and logwood in every dye-house where they could be found, and this act remained in full force till the time of Charles the Second, a period embracing a considerable part of a century. A foreigner might have supposed that the legislators of England entertained such an affection for their native woad, with which their denuded sires used to stain their skins in the olden times, that they would allow no outlandish drug to come in competition with it. A most instructive and interesting volume might be written, illustrative of the evils inflicted upon arts, manufactures, and commerce, in consequence of the ignorance of lawgivers.

When these absurd prejudices were gradually overcome in the eighteenth century, the art of dyeing made considerable progress. Madder, from which the color known as Turkey or Adrianople red is produced, then began to be properly appreciated; and quercitron, a fine yellow dye, was brought extensively into notice by Dr. Bancroft. But the chief improvements of the moderns in this art, consist in the employment of pure mordants, and in the application of colors derived from mineral compounds, as sesquioxide of iron, prussian-blue, chrome-yellow, chrome-orange, manganese-brown, etc. Each of these may be obtained as an insoluble precipitate, by mixing together two dissolved salts; in the dyeing processes, the proper solutions are made to mingle, and produce the deposit within the fibre by impregnating first with one solution and afterwards with the other. As the precipitate thus produced is imprisoned within the fibre, it is not removable by mere aspersion with water.

In India, was discovered the mode of dyeing Turkey red, which is the most durable vegetal tint known. It was afterwards practised in other parts of Asia and in Greece; and about the middle of last century, dye-works for this color were established near Rouen and in Languedoc by several Greeks. In 1765 the French government, convinced of the importance of the process, caused an account of it to be published; but it was not introduced into England until the end of the eighteenth century, when a Turkey-red dye-house was established in Manchester by M. Borelle, who obtained a grant from government for the disclosure of his process. The method, which was made public, does not seem to have been very successful. A better mode was introduced into Glasgow about the same time by another Frenchman, named Papillon.. Previous to this, however, Mr. Wilson of Ainsworth, near Manchester, had obtained the secret from the Greeks at Smyrna, which he revealed; but the process was said to be expensive, tedious, and less applicable to manufactured goods than to cotton in the skein. The greater part of the Turkey-red dyeing executed in Great Britain, is still carried on in Glasgow.

CHAPTER II.

CHEMICAL PRINCIPLES OF THE ART OF DYEING.

The art of dyeing has been of late so scientifically cultivated that it would require a greater space than the limits of this treatise can afford, to give a complete idea of it, and we shall confine ourselves to the explanations of the chemical principles, on which are based the preliminary preparations of the textile fibres to render them fitted for the manufacture of tissue and those on which is founded the art of fastening coloring matters.

Preparation of the Textile Fibres.

The textile fibres used in manufactures are either of vegetable or animal origin; the first being chiefly *Hemp*, *Flax*, and *Cotton*, and the second *wool*, *hair* of animals, and SILK spun by the silk worm.

Cotton is nearly pure lignin, while hemp and flax are composed of lignin in long filaments, which, when dry, adhere to each other by means of a gelatinous substance called *Pectin*, although it differs probably from that found in fruits, and which must be removed to render them fit for spinning and weaving. For this purpose they are *rotted*, which operation consists in plunging them tied in bundles, into water, where they are left, until fermentation commences, which is manifested in stagnant waters, by a very disagreeable odor; the bundles are then withdrawn from the *rotting pond*, and, after having been dried in the air, are subjected to a mechanical operation of which the object is to detach the foreign substances, which have become friable by the desiccation ensuing on the rotting, and to isolate the fibres. Hemp and flax thus prepared are fit to be connected by spinning into *unbleached thread*, which may be immediately used for weaving cotton, undergoes no preliminary preparations, and may be immediately spun and woven.

Wool, as it is found on the living animal, is impregnated with a considerable quantity of foreign matters, commonly called *grease* (SUINT), and which consists essentially of substances soluble in water, and fatty substances insoluble in that fluid. Sheep are usually washed before being shorn, and then yield what is called *washed wool*, which has just lost a large portion of its soluble matters, and a portion of the fatty matters, which separated in the state of an emulsion. Wool which has not undergone this operation is called *unwashed wool*, and the process by which the grease is removed from wool is known by the name of *scouring*. Unwashed is scoured with wash wool in a bath of 84 gallons of water, and 20 to 22 gallons of putrefied urine, the whole being heated at 122° or 140° for *soft wool*, and to 158° or 167° for *harsh wool;* after dipping 6 lbs. 12 oz. or 9 lbs. of unwashed wool into the bath, and stirring it with a stick for 10 minutes, they are removed and allowed to drain over the kettle, the same being done with another lot, until about 90 lbs. in all have been thus treated; $1\frac{1}{2}$ galls. to 2 galls. of putrid urine are then added, and 112 lbs. of washed wool passed through it, which is scoured both by the carbonate of ammonia of the putrefied urine and the alkaline substance yielded by the unwashed wool. The same operation is repeated on a new lot of 90 lbs. of washed wool, after which a new dose of $1\frac{1}{2}$ to 2 galls. of putrid urine is added, and 45 lbs. of unwashed wool, washed in it. This alternate scouring of wash and unwashed wool is continued during the whole day, adding urine at each fresh quantity of unwashed wool. After this operation the unwashed wool should be considered as wash, and treated accordingly.

When the wool scourer has no unwashed wool, he makes his bath of 183 galls. of water and 84 galls. of urine, heats it at 120° or 140° and passes through it 68 lbs. of wool in 5 lots, each of which he leaves in the bath for 12 or 15 minutes, after which he adds 2 pints of water and $\frac{1}{2}$ gall. of urine, and then scours an additional portion of 68 lbs. of wool, &c. Some scourers add marly clay to the bath.

Wash wool contains less than 15 per cent. of grease, while unwashed contains much more, and by washing, scouring, and drying loses as much as 60 or 70 per cent. of its weight. When the washed wool contains less than 5 per cent. of grease, it is scoured with soap or carbonate of soda.

The nature of the fatty matters of the grease is peculiar, and they have been called by Mr. Chevreul *stearerin* and *elaierin;* the first is solid, but uncrystallizable; the second is oleaginous. These fats are not saponified by weak alkalies, but when they are boiled for a long time with a solution of caustic potash, the liquid is found to contain two salts of potash, formed by peculiar fat acids which have been called *steareric* and *elaieric acids*, while nothing analogous to glycerin has been found, the oxygen of the air may possibly have some share in the formation of these fat acids.

After scouring, the wool is washed in river water, in willow baskets. When it is intended to be perfectly white, it is exposed for some time in a moist state in rooms in which sulphur is burned, where the sulphurous acid finishes the bleaching, and the excess of it is removed by fresh washings. It is important not to prolong too much the action of the sulphurous acid, because it exerts a decomposing agency on the nitrogenous substance of the wool.

Wool contains a proximate sulphuretted principle, which may be separated by successive immersions in lime-water. Wool which has been heated with a weak alkaline solution, disengages sulphydric acid, when it is again heated with acidulated water, and is blackened when boiled with a solution of a salt of lead or protoxide of tin.

Raw Silk, as obtained from the cocoons, is impregnated with a gelatinous substance, which makes it very stiff, and generally gives it a golden yellow tinge. This substance, which forms about $\frac{1}{4}$th of the weight of raw silk, dissolves readily in alkaline liquids, but as caustic alkalies attack the silk itself, soap is almost always used, and sometimes, but rarely, carbonate of soda.

The operation which is called *Scouring* (DECREUSAGE) *the silk*, is divided into three stages, the *ungumming* (DEGOMMAGE), *boiling*, and *bleaching*. The ungumming is done in a tin boiler containing for every 100 parts of silk, 1800 or 2500 parts of water, and 30 of soap. It is first boiled to dissolve the soap, and then cold water is added so as to lower the temperature at about 200°, when the silk is dipped into it in skeins, supported by sticks called *lisoirs*, being there left until all the gelatinous matter is dissolved, and afterwards wound on a bobbin. This operation lasts from $\frac{3}{4}$ to $1\frac{1}{2}$ hours. Several skeins are then united, forming a *hank*, which is boiled for $1\frac{1}{2}$ hours in a bath con-

4

taining 20 or 30 parts of soap for 2000 parts of water, which constitute the *boiling* (CUITE). The hanks are undone, twisted into skeins, wound on a bobbin, and then washed in a weak solution of carbonate of soda, and in water. The bleaching consists in dipping the silk held by the lisoirs, into a bath heated at 203°, and composed of 84 galls. of water, and from 1 lb. 2 oz. to 1 lb. 12 oz. of white Marseilles soap. Silks which are intended to be white, are exposed in addition to sulphurous acid.

CHAPTER III.

PRELIMINARY PREPARATION OF STUFFS.

Before being printed, cotton stuffs are *singed* with the intention of removing the filaments which project from the tissue. The shearing is performed by machines called *shearing machines*, composed of two revolving cylinders, one of which, furnished with brushes, raises the nap, while the other, provided with knives arranged spirally, shears it. In singing, the stuff is passed rapidly over a metallic cylinder, heated to nearly a white heat, which burns off the down. Cotton stuffs intended to be perfectly white, are previously *bleached*, which operation is also more or less completely performed on goods which are to be printed.

Linen and cotton goods are bleached by two processes: 1. By washing them in alkaline lyes, and exposing them on the grass. 2. By chlorine and by the alkaline hypochlorites.

The first is the oldest, and was used particularly for bleaching flax and hemp goods. It is divided into the following operations: 1. *Scouring*, which consists in dipping the stuffs for twenty-four hours in a weak solution of caustic potash, heated at about 99°, washing, and then boiling them for twenty minutes in the same alkaline lye.

2. The *boiling*, which consists in boiling the scoured stuffs, after having washed them in water, and compressed them between cylinders.

3. *Bleaching*, which consists in boiling them for six hours with an alkaline lye containing 1 part of caustic potash for 16 parts of stuff, washing them, and exposing them for five or six hours on the grass; the alkaline washings and exposure on the grass being renewed until the stuffs are perfectly bleached. During the exposure on the grass, the coloring matters are bleached by the influence of the solar rays and moisture; the absorption of oxygen converting them into new substances, more readily soluble in the alkaline liquors. Lastly, the stuffs are passed through water heated at 105° or 120°, containing about $\frac{1}{50}$ of sulphuric acid, which dissolves the metallic oxides, after which they are washed and calendered.

This process requires a great length of time, and bleaching by the hypochlorites or chlorine is more expeditious. The chlorine acting on the coloring matter in the presence of the water, decomposes this water into hydrogen and oxygen; hydrogen combines with the chlorine to form hydrochloric acid, while oxygen in the nascent state oxidizes the resinous and coloring matters, and renders them soluble in alkaline lyes. The hypochlorites are reduced to the state of chlorides, and act at the same time by means of the nascent oxygen given off by the hypochlorous acid and the base, while the concurrence of an acid effecting the decomposition of the hypochlorites hastens the bleaching. Thus in both processes it is in the end always an oxidizing action, which effects the bleaching and destruction of the foreign substances.

Hypochlorite of lime, dissolved in water, is now solely used in bleaching, and it is preferable to all dilute solutions, because it is less liable to injure the ligneous fibre of the tissue, although the bleaching then requires more time.

The stuffs, after being passed over the heated cylinder to be singed, are immediately dipped into a vat filled with water to cool them, where they then remain for twenty-four hours, and lose a considerable portion of their soluble principles. They are then to be perfectly dried, either by being beaten or compressed between cylinders, and then kept for twelve hours in a vat filled with water heated by steam, where they are arranged in alternate layers with slaked lime; after being again beaten, they are left for twelve hours in a lye of caustic soda, consisting for 300 parts of stuffs, of 10 parts of caustic soda for 1500 of water. This lye is replaced by another containing only 7.5 of soda, which is also allowed to act for twelve hours; after which the stuffs, pressed dry, are passed through the hypochlorite of lime, and then through sulphuric acid. The bath of hypochlorite generally contains 0.15 parts of chlorine or a quart of water; and the stuffs after being immersed in it are passed between two wooden cylinders, descending them immediately into a bath acidulated with sulphuric or hydrochloric acid, which hastens the bleaching by isolating the hypochlorous acid.

After being washed in fresh water, they are for a second time subjected to the action of alkaline lyes, hypochloride of lime, and the acid baths, and lastly, after another washing in fresh water, they are dried in washing machines, and more body is given to them by dressing them with starch.

CHAPTER IV.

MORDANTS.

THE tissues of muslin or linen stuffs have, for a great number of coloring substances, an affinity sufficiently powerful to fasten them on their surfaces, and to acquire a deep color, while the combination is nearly strong enough to enable them to resist washing, particularly with alkaline soaps. They are made fast, and at the same time the color is heightened by previously depositing on the tissues certain substances which have a greater affinity for these tissues than the coloring matter, and which possess, at the same time, the property of forming, with the coloring matters, compounds sufficiently fixed to resist washing in fresh water and in soapsuds. These substances which thus play an intermediate part between the woven fabrics and the coloring matters, are called *mordants*. The affinities, by virtue of which they are fastened on the fabric, exhibit this essential difference from those observed in ordinary chemical operations, that, in the latter, combination generally ensues only between disaggregated substances, and if one of the substances is originally aggregated, it becomes disaggregated by the simple fact of combination; while, in dyeing, the woven fabric retains its form and consistence, without being in the slightest degree disaggregated by the mordants and coloring matters. Certain mordants do not change the shade of the coloring matters, such, for example, as the salts of alumina and chloride of tin; while others, on the contrary, alter the color, as the salts of iron, copper, manganese. The salts of alumina, used as mordants, are the sulphate and acetate of alumina and alum; the fastening of color by alum being called *aluming*.

In order to alum cotton, flax, or hempen stuffs, they are left for twenty-four hours in a tepid bath, containing one pound of alum for six pounds of fabric, when a portion of the alum adhering to the stuff, renders the latter fit for dyeing. For dark colors, the ordinary commercial alum is used; pure alum being preferred for bright colors, because common alum contains a small quantity of sulphate of iron, which would modify the color.

Wool is alumed by being first boiled in bran-water for an hour, and washed in fresh water, and then kept for two hours in a boiling solution which contains ten to fifteen per cent. of alum, a small quantity of cream of tartar being generally added, which facilitates the deposit of alumina on the tissue, probably in converting a portion of the sulphate of alumina into a tartrate more easy to decompose. When the wool is alumed, it is left for two days to rest before dyeing, in order to render the combination of the mordant with the fibre more intimate.

Silk is alumed when cold, by keeping it for fifteen or sixteen hours in a bath containing $\frac{1}{80}$ of alum, after which it is removed and washed. Acetate of alumina, which is often used as a mordant for ligneous stuffs, and for certain colors, is prepared like we shall see hereafter, by decomposing alum by acetate of lead. The solution of acetate of alumina thus obtained being generally thickened with gum or starch.

Stuffs of lignin, mordanted with alum, are again subjected, before being dyed, to another operation, the effect of which is not well understood; they are immersed for some time in two baths of water, containing from six to eight per cent. of cow-dung. To the first of these baths a certain quantity of chalk is added, the intention of which appears to be to saturate the acid partly adhering to the tissue with the mordant; while the second contains only water and dung. The temperature of these two baths varies according to the nature of the stuffs and that of the mordants. The cow-dung appears to act by means of the phosphates it contains, for a mixture of phosphate of soda and lime can be substituted for it.

Protochloride of tin is chiefly used for obtaining the oxide of tin as a mordant, which adheres very firmly to the tissues. Bichloride of tin is often used for freshing colors, particularly those of cochineal and madder.

The mordant of oxide of iron is furnished by the proto-acetate, prepared by the action of pyroligneous acid on old iron.

The question of mordants is so important, that we will treat it hereafter at some length.

CHAPTER V.

DYEING.

AFTER the stuffs are mordanted, they are immersed in order to be dyed, in solutions of coloring matters of various temperatures, and then left for a longer or shorter time, according to the nature of the stuff and the tint of color to be obtained. It is essential that all parts of the fabric should remain the same length of time in the dye; to which effect it is rolled around a wooden roller suspended under the dye tub, and is unrolled through the tub, this process being continued until the color has obtained the shade required. In order to obtain a regular shade, it is better to use successive baths of different strength, commencing with the weakest. The baths are sometimes composed of a single coloring matter, and sometimes of a mixture of several, while at other times the stuff is passed successively through two baths containing different colors, and thus an intermediate shade is obtained; the colors are fastened by washing in soapsuds or in other solutions.

It would lead us too far to give a description of the methods of preparing the different solutions for dyeing and the manipulations of the process. For this we refer the reader to a regular work on the art of dyeing.

CHAPTER VI.

ON THE COLORING MATTERS PRODUCED BY COAL TAR.

History.—Until the year 1854, ANILINE was known only by chemists; it was a product of the laboratory which was found only with difficulty; still, Industry had not the less desire to use it, on account of its high price and its difficult and costly preparation. At that time, Mr. Dumas presented to the Academy of Science of Paris a paper *on a new method of formation of artificial organic bases,* in which Mr. Bechamp made known a process by which he was enabled to obtain *Aniline* not only easily, but also at a low price.

By the efforts of Messrs. Renard Brothers, Franc, Tabourin and Bechamp, Aniline is now a product which can be obtained easily.

In 1826, Unverdorben, studying the products which result from the dry distillation of animal matters with indigo, discovered among the pyrogeneous products of this last substance, an organic basis, volatile, liquid, and heavier than water, which he called *Crystalline,* because with mineral acids it produces easily crystallized salts.

Mr. Fritzsche afterwards studied these products, and called Aniline (from the name of the *indigofera anil*) the basis obtained in distilling indigo with caustic potash. He demonstrated that this basis was identical to Crystalline. Subsequently Mr. Runge isolated, by a process modified by Hoffmann, among the bases which exist in the heavy oils of the distillation of coal tar, an oily organic basis from which he developed a fine violet blue color by hypochlorite of lime.

Mr. Zinin afterwards, by the action of sulphuret hydrogen on nitro-benzine in connection with ammonia, produced an organic basis which he called *Benzidum*.

When the identity of all these products was established, chemists adopted the name of *Aniline* to designate them all, this title being the best for the formation of compound names.

These first experiments gave birth to others which showed that in a multitude of reactions, Aniline could be produced, so it could be formed by the action of alkalies and alcohol on nitro-benzine.

Messrs. Laurent and Hoffmann, in heating for fifteen days, in a tube, *phœnic acid* with ammonia, have also produced Aniline.

During all the time that this product could be made only by the above processes, Aniline was simply an object of curiosity. Its extraction from

COLORING MATTERS PRODUCED BY COAL TAR.

coal tar was difficult, and from indigo, the quantity produced was too small and too costly.

Mr. Perkins, the great English manufacturer, studied the production of Aniline at the same time as several French chemists, but the French being too much engaged with the theoretical question, left to Mr. Perkins the honor of the industrial discovery. It was with the benzine (benzole) that he succeeded in producing the largest quantities of Aniline.

CHAPTER VII.

DISTILLATION OF COAL TAR.

The dry distillation of organic matters, vegetable or animal, from the great variety of products to which it gives rise, constitutes one of the most interesting operations of chemistry.

Their reactions are very complex, and some of them have been very little studied, as indeed is the case with many of the substances formed.

If the body submitted to dry distillation could be maintained during the operation under uniform conditions of desiccation, temperature, and pressure, the reactions and the products would be more simple. If, for example, wood be heated very slowly in close vessels, first to 212° F., then to 392° and 572°, and so on, there is at first disengaged almost pure water, then impure strong acetic acid, and afterwards a mixture of acetone and acetate of methylene; the maximum of charcoal is left as residue, and the least amount of tar and gas is produced, the latter consisting only of carbonic acid and carburetted hydrogen.

In practice, however, when wood is distilled in iron cylinders, heated from the outside, the heat only penetrates to the interior gradually. The outside layers are, therefore, the first decomposed; they at first lose water, then furnish pyroligneous acid and wood spirit, at the same time giving off carbonic acid and a little carburetted hydrogen.

The inner layers in turn are similarly decomposed, but the products as they are given off are brought into contact with the outer layer, already in a more advanced state of decomposition and at a much higher temperature, and hence new reactions take place and new products are formed. Thus, the vapor of water in contact with red hot charcoal is decomposed, and forms carbonic acid and hydrogen; a part of the carbonic acid is again decomposed by the red hot carbon to form some oxide of carbon. A part of the nascent hydrogen combines with carbon to form various hydrocarbons; one part of the acetic acid is decomposed by the high temperature to form acetone and carbonic acid; another part reacts on the wood spirit, and forms methylic acetate; a fraction of the wood spirit and acetone are also decomposed, producing *tarry matters, pyroxanthine, oxyphenic acid, dumasine*, etc. To these must be added the influence of certain nitrogenized bodies, and we can understand how all these compounds, successively formed under the most favorable circumstances for acting on one another, since they are in the nascent state and exposed to a high temperature, may give rise to the formation of a great variety of different compounds, which will be set free either in the state of a permanent gas or of a condensable vapor, and leave fixed carbon as a residue.

The same takes place whether wood, coal, asphalte, peat, resin, oils, or animal matters be distilled; but it is evident that the original composition of the material submitted to dry distillation must powerfully influence the nature and composition of the products. In those which, like wood, are rich in oxygen and poor in nitrogen, the pyrogeneous products contain much acetic acid and but little ammonia, and consequently have an acid reaction; on the contrary, the matters containing much nitrogen, and but little oxygen, like coal and animal matters, give rise to the formation of much ammonia, and the products have an alkaline reaction.

In this division we intend only to confine our attention to the products obtained by the distillation of coal tar from gas works. Considerable differences are noticed in the composition of the tar procured from different qualities of coal and schists, according to the rapidity with which the distillation has been conducted. Some tars, for instance, contain but little benzole, but much naphthaline; boghead tar is rich in paraffine; others contain a preponderating quantity of phenyl and benzole.

DISTILLATION OF COAL TAR.

Table of the Products Obtained by Distillation and Rectification of Coal Tar.

SOLID PRODUCTS.

Carbon,	or Anthraceine,	Chrysene,
Naphthaline,	Paraffine,	Pyrene.
Paranaphthaline		

LIQUID PRODUCTS.

Acids.	Neutrals.	Bases.
Rosolic,	Water,	Ammonia,
Brunolic,	Essence of Tar,	Methylamine,
Phenic,	Light Oil of Tar,	Ethylamine,
Phenol,	Heavy Oil of Tar,	Aniline,
Acetic,	Benzole,	Quinoline,
Buthyric.	Toluole,	Picoline,
	Cumole,	Toluidine,
	Cymole,	Lutidine,
	Propyle,	Cumidine,
	Butyle,	Pyrrhol,
	Amyle,	Poetinine.
	Caproyle,	
	Heptylene,	
	Hexylene.	

GASEOUS PRODUCTS.

Hydrogen,	Various Hydro-Carbides,	Carbonic Acid,
Carburetted Hydrogen,		Sulphydric Acid,
	Oxide of Carbon,	Hydrocyanic Acid.
Bicarburetted Hydrogen,	Sulphuret of Carbon,	

Whatever may be the composition of the different kinds of tar, they are all submitted to distillation in order to isolate the principles capable of industrial application. But, first of all, it is necessary to separate the tar, as far as possible, from the ammoniacal liquor which is found with it. For this purpose, it is heated some hours at 176° or 212° F., by which it is rendered more liquid, and then the water separates more easily. It is then allowed to cool very slowly, and the water is drawn off by a tap placed at the lower part of the boiler. A certain quantity of tar obstinately retains the water, constituting a buttery matter, which may be allowed to run away with the water, to be added afterwards to another quantity of tar to be deshydrated by a fresh operation.

Experience seems to have demonstrated that the most simple process, that is to say, distillation over a naked fire at the ordinary pressure, is still the most practicable and advantageous. As the volatile products have but little latent heat, the height of the still should be somewhat less than the diameter; for the same reason the head must be carefully protected from cold, and it is well to furnish the inside with a circular gutter, in which the products condensed in the head may be collected and run into the refrigerator. By this means the products are prevented from flowing back into the boiling tar, and being decomposed by coming in contact with the sides of the still, which, especially towards the end of the operation, becomes very hot.

In condensing the vapors, it is necessary to

observe certain precautions. At the beginning of the operation, when the lighter and more volatile oils are passing, the worm must be well cooled to make quite sure of the condensation. Later, when the heavier and less volatile products are coming over, the water in the refrigerator may be allowed to get heated at 86° or 104° F., and at last when the matters capable of solidifying, such as naphthaline and paraffine, pass, the temperature of the refrigerator should never be under 104° F., and it may be allowed without inconvenience to raise to 140° or 158° F. At this temperature the products condense perfectly, but remain liquid and run with ease. If the refrigerator was kept quite cold during the whole process, it might happen toward the end, that the condensed tube would become blocked up by the solidified products, and a dangerous explosion might ensue.

At the beginning of the distillation the tar should not be allowed to boil too fast. Some distillers at this period pass a current of steam at 230° or 248° F., through the tar to assist the disengagement of the more volatile oils.

These, in condensing, form a very limpid fluid liquid, having the density of .780, which gradually rise to .850; the mean density of all the products united is about .830. It is this, which constitutes the benzine of commerce. It contains a great variety of compounds whose boiling points range from 140° to 392°. They belong to the following series:—

$C^n H^n$ e. g.	Amylene, $C^5 H^5$	
	Hexylene (oleine Caproylene)	$C^6 H^4$
	Hepthylene (Oenenthylene)	$C^7 H^7$
		etc.
$C^n H^n + 2$ e. g.	Propyle	$C^{12} H^{14}$
	Butyle	$C^{16} H^8$
	Amyle	$C^{20} H^{22}$
		etc.
$C^n H^n - 6$ e. g.	Benzine	$C^{12} H^6$
		etc.

When the density of the products exceeds .850, the current of steam is stopped and the heat increased. As soon as the temperature of the tar has risen from 392° to 428° F., the distillation recommences, and the oil condensed is found to have a sp. gr. .860 to .900, the mean being from .880 to .885. This product constitutes the heavy oil of tar, and contains phenol, creasote, and aniline.

Lastly, the ultimate products of the distillation, which on cooling become a buttery mass, or crystalline, if they contain much naphthaline, are set aside for the preparation of paraffine. They are placed in vats, which are cooled, in order that the solid matters may separate by crystallization.

2000 parts of rough oil of tar obtained by the distillation of Boghead coal furnished on rectification:—

DISTILLATION OF COAL TAR.

```
1208 parts light oil, density =     .825
 200   "   heavy oil =   .    .   .860
 400   "   pitch
 192   "   gas escaped
```

2900 parts of tar from gas works using Boghead coal, distilled in a similar manner, yielded:—

```
Water, slightly ammoniacal    .   .   .   .  168
Light hydro-carbons, mean density .820 .   .  480
Heavy hydro-carbons, mean density, .863   .  883
Fatty pitch, solid when cold, liquid at 302° F. 1195
Loss 6 per cent.   .   .   .   .   .   .  174
                                         ─────
                                         2900
```

CHAPTER VIII.

HISTORY OF ANILINE—PROPERTIES OF ANILINE—PREPARATION OF ANILINE DIRECTLY FROM COAL TAR.

§ 1. *History of Aniline.*

ANILINE was discovered in 1826 by Unverdorben. The original method for its preparation was by digesting indigo with hydrate of potash, and subjecting the resulting product to distillation. Aniline was also obtained from the basic oils of coal tar; but the process which is now employed for its preparation is a remarkable instance of the manner in which abstract scientific research becomes, in the course of time, of the most important practical service. It was Faraday who first discovered benzole; he found it in oil-gas. After this it was obtained by distilling benzoic acid with baryta, which result determined its formula, and was the cause of its being called *benzole*. After this, Mansfield found it to exist in large quantities in common coal tar naphtha, which is the source from which it is now obtained in very large quantities. Benzole, when studied in the laboratory, was found to yield, under the influence of nitric acid, nitro-benzole. Zinin afterwards discovered the remarkable reaction which sulphide of ammonium exerts upon nitro-benzole, converting it into aniline. And, lastly, Bechamp found that nitro-benzole was converted into aniline when submitted to the action of ferrous acetate. It is Bechamp's process which is now employed for the preparation of aniline by the tun. Had it not been for the investigations briefly cited above, the beautiful aniline colors now so extensively employed, would still remain unknown. When Mr. Perkins discovered aniline purple, nitro-benzole and aniline were only to be met with in the laboratory; in fact, half a pound of aniline was then esteemed quite a treasure, and it was not until a great deal of time and money had been expended that he succeeded in obtaining this substance in large quantities, and at a price sufficiently low for commercial purposes.

The coloring matters obtained from aniline are numerous; they are the following: Aniline purple, violine, roseine, futschine, alpha aniline purple, bleu de Paris, nitroso-phenyline dinitraniline, and nitro-phenyline diamine.

§ 2. *Chemical Properties of Aniline.*

Pure aniline is a colorless liquid, very astringent, having an aromatic odor and an acid burning taste, slightly soluble in water, very soluble in alcohol and ether.

Its specific gravity = 1.028. It does not freeze at —20°.

It boils at 262°.4 F., and distils unchanged. When warmed it dissolves sulphur and phosphorus.

It is a powerful basis, combining with acids, and forming salts, which in general are soluble.

It decomposes salts of protoxide and peroxide of iron, and the salts of zinc and alumina, precipitating from them the metallic oxides.

It precipitates also the chlorides of mercury, platinum, gold, and palladium, but does not precipitate the nitrates of mercury and silver.

Aniline easily oxidizes, turning yellow in water, and in time becoming resinified.

When aniline dissolved in hydrochloric acid is acted on by chlorine, the solution takes a violet color, and on continuing the current of chlorine, the liquid becomes turbid and deposits a brown-colored resinoid mass. In distilling the whole, vapors of *trichloraniline* and *trichlorophenic acid* pass over.

A solution of the alkaline hypochlorites colors aniline violet blue, which turns rapidly red, especially in contact with acids.

A mixture of hydrochloric acid and chlorate of potash acts on aniline, the final result of the action being *chloranile* $C^{12} Cl^4 O^4$, but in the course of the reaction several colored intermediary bodies are formed.

If a solution of chlorate of potash in hydrochloric acid be added to a solution of a salt of aniline mixed with an equal volume of alcohol, and care is taken to avoid an excess of the hydrochloric solution, a flocculent precipitate is deposited after a time of a beautiful indigo blue color; this precipitate filtered and washed with alcohol contracts strongly, and passes to a deep green. The filtered liquid has a brownish red color; on boiling it, adding fresh quantities of hydrochloric acid and chlorate of potash, a yellow liquor is obtained, which deposits crystallized scales of *chloranile*.

An aqueous solution of chromic acid gives, with solutions of aniline, a green, blue, or black precipitate, according to the concentration of the liquors.

When a small quantity of an aniline salt is mixed in a porcelain dish with a few drops of strong sulphuric acid, and a drop of a solution of bichromate of potash is allowed to fall on the mixture, a beautiful blue color appears after some minutes, which, however, soon disappears.

Diluted nitric acid combines with aniline without adhering to it immediately; but after some time nitrate of aniline crystallizes in the form of concentric needles, the mother liquor turns red colored, and the sides of the evaporating dish become covered with a beautiful blue effervescence. When a few drops of strong nitric acid are poured upon aniline, it is immediately colored a deep blue; on applying heat the blue tint quickly passes to yellow, a lively reaction is manifested, which results in the formation of *picric acid*, or *trinitrophenisic acid*.

Potassium dissolves in aniline, disengaging hydrogen, whilst all becomes a velvet-colored pap.

The other reactions of aniline which are characterized by the formation of Futschine Azalcine, will be related in the sequel of this book, when describing their preparations.

§. 3. *Preparation of Aniline directly from Coal Tar.*

The method which appears to be the most rational, and which deserves to be tried, would consist in treating the tar as condensed in gas works with hydrochloric or sulphuric acid, diluted with three or four times its volume of water. Mechanical means for affecting the intimate mixture of the tar with the acid might be easily contrived, but in the absence of any special contrivance, the end may be obtained by half filling a barrel with the tar, adding one-fifth or one-sixth of its volume of acid, and rolling and shaking the barrel until the acid has taken up the bodies with which it is able to combine; the whole might thus be run into a cistern, where, by degrees, the watery liquid would separate from the tar.

The same acid liquid might be used over and over again until the bases have nearly saturated the acid. A very impure aqueous solution would thus be obtained, containing the hydrochlorates or sulphates of ammonia, and all the other organic bases contained in the tar, such as *aniline, quinoline, pyrrol, picoline, pyrrhidine, lutidine, toluidine, cumidine,* etc.

By evaporating this solution almost to dryness, and then distilling with an excess of milk of lime, the bases would be set at liberty. Ammonia, as the most volatile, would be disengaged first, and might be condensed apart, and by raising the temperature higher and higher, the organic bases would be disengaged. Aniline would be found among the liquids distilling between 302° and 482° F.

The manipulation of the tar, however, is an extremely disagreeable operation, and presents many difficulties; it is therefore preferable, in many cases, to distil the tar first, and operate on the most pure and limpid distilled oil.

Aniline, because of its high boiling point, is never met with, in the light and volatile liquids when first distilled from tar. The most of it is found in those which distil between 302 and 356.° These, according to Hoffmann, contain about 10 per cent. of organic bases, mostly *aniline* and *quinoline.* The oils which distil above 482°, contain mostly quinoline and very little aniline.

The following process for extracting the two bases from the oil and separating them, is due to Hoffmann. The oil is agitated strongly with commercial hydrochloric acid. The mixture is then allowed to rest for 12 or 14 hours, and the oil is separated from the acid; the latter is treated again by fresh quantities of oil until nearly saturated. The still acid solution is filtered to retain the oil interposed mechanically. It is then placed in a copper still and supersaturated with an excess of milk of lime. At the moment of saturation an abundance of vapors are given off, and the head must be quickly fixed on the still. Heat is now applied so as to obtain a quick and regular ebullition.

The condensed product is a milky liquid with oily drops floating on it. The distillation is carried on, as long as the vapor has the peculiar odor of the first part distilled, or the condensed product gives the characteristic reaction of aniline with chloride of lime.

The milky liquid is now saturated with hydrochloric acid; it is then concentrated in a water bath; and lastly, decomposed in a tall narrow vessel by means of a slight excess of hydrate of potash or soda. The bases set free, unite and form an oily liquid, which floats on the alkaline solution. This is removed with a pipette and rectified. The rectified product is aniline, sufficiently pure for industrial purposes, especially if we set aside the part distilling above 392° or 428° F., which is principally composed of quinoline.

To obtain aniline chemically pure, the neutral

oils forming part of the oily layer must be completely removed. This is done by dissolving the whole in ether, and adding dilute hydrochloric acid, which combines with and separates the bases, and leaves the oil in solution in ether. The acid solution is then decanted, decomposed with potash, and submitted to careful fractional distillation. If the products are gathered separately in three parts, the first will contain ammonia, water, and some aniline; the second will be pure aniline; while the third portion will contain mostly quinoline. An alcoholic solution of oxalic acid is now added to the impure aniline, which precipitates *oxalate of aniline*, as a mass of white crystals, which are washed with alcohol, and then pressed. The salt is then dissolved in a small quantity of water, to which a little alcohol is added. From this solution, the oxalate crystallizes in stellated groups of oblique rhomboidal prisms. These crystals are decomposed by a caustic alkali, to set free the aniline, and when this is distilled, water at first passes, then water charged with aniline, and lastly, at 359° F., chemically pure aniline.

CHAPTER IX.

ARTIFICIAL PREPARATION OF ANILINE—PREPARATION OF BENZOLE—PROPERTIES OF BENZOLE—PREPARATION OF NITRO-BENZOLE—TRANSFORMATION OF NITRO-BENZOLE INTO ANILINE, BY MEANS OF SULPHIDE OF AMMONIUM; BY NASCENT HYDROGEN; BY ACETATE OF IRON; AND BY ARSENITE OF POTASH—PROPERTIES OF THE BI-NITRO-BENZOLE.

Artificial Preparation of Aniline.

THIS process constitutes one of the most important and curious reactions of organic chemistry; it enables us to obtain aniline in any quantity. It is not difficult to prepare, but certain precautions are however necessary, when operating on a large scale. The process can be subdivided into three distinct operations:—

1. Preparation of benzole.
2. Transformation of benzole into nitro-benzole.
3. Reduction of nitro-benzole into aniline.

§ 1. *Preparation of Benzole.*

The only process we think necessary to notice is that by which benzole is obtained on a large scale, viz: the extraction from coal tar, or from the first products of the distillation of coal tar, light oil, or crude naphtha.

The manufacturer who wishes to distil tar in order to procure the largest amount of benzole, should choose a light fluid tar, and especially one distilled from boghead or cannel coal. To form a comparative estimate of the value of different tars, the following experiment may be performed:—

About $2\frac{1}{2}$ gls. of tar are distilled until the vapors, instead of condensing into a liquid, furnish a product which, on cooling, becomes solid, or of a buttery consistence. By carefully observing when the condensed oil becomes heavier than the water, and measuring the volume of the lighter oils which float on the surface of the water, and then comparing the volumes, we are enabled to estimate with tolerable accuracy the value of the tar. Of course, the one which yields the largest amount of light oil is the best.

Crude naphtha, or the benzole of commerce, is generally a yellow or brown liquid, having a density varying from .90 to .95; it usually contains, besides benzole, some of the homologues of benzole, toluol, cumol, and cymol. It is impossible to separate these bodies by an ordinary process of rectification; for although the boiling point of tolrol is 226° or 228°, and that of cumol 289° or 293°, their vapors are, so to say, dissolved in the vapor of benzole, and are carried over and condensed together. Their presence, however, does not interfere with the preparation of nitrobenzole and aniline.

When you have obtained the light oil from the coal tar, wash it with a little sulphuric acid (10 per cent. of strong acid). Leave it one hour, and saturate with soda.

Distil; the product escapes through a cool worm.

In the receiver are two oils, one lighter and the other heavier than water, the first occupies about one-tenth of the total volume: it is the benzole; add to it a little sulphuric acid, wash and distil it.

The benzole found in commerce is sometimes very impure; some has been met with, containing merely a trace of real benzole. Such an article is ordinarily the result of the distillation of bituminous schists or asphaltum, and besides hydrocarbons belonging to another series than that of benzole, it generally contains a small amount of oxygenated products, and consequently cannot be advantageously used in the preparation of aniline. It is therefore important to be able to detect benzole in a mixture of other oils. For this purpose we may avail ourselves of the facility with which true benzole is converted into nitro-benzole, and then into aniline by the action of nascent hydrogen.

The following is Hoffmann's method: a drop of benzole is heated in a small test tube, with fuming nitric acid, to convert it into nitro-benzole. A good deal of water is then added, to precipitate the nitro-benzole in small drops, which must be taken up by ether. The ethereal solution is then poured into another small tube, and equal volumes of alcohol and diluted hydrochloric acid are then added; a few fragments of granulated zinc are then dropped in. In about 5 minutes sufficient hydrogen will have been disengaged to produce aniline, which will be found combined with the acid. The liquid is supersaturated with an alkali and shaken with ether, which dissolves the aniline set free. A drop of this ethereal solution allowed to evaporate in a watch glass, and mixed after the evaporation of the ether with a drop of a solution of hypochlorite of lime, will show the violet tints which characterize aniline. The operations may be executed rapidly, and without any difficulty.

Properties of Benzole.

At the ordinary temperature, benzole is in the form of a colorless, very fluid liquid, of an agreeable odor, and has a specific gravity of .85 at

59° F. At a very low temperature it crystallizes or forms a mass like camphor, which melts at 41°.

Its boiling point is between 176° and 170°.8; and it distils without undergoing any change. It is nearly insoluble in water, to which it imparts its peculiar odor; it is very soluble in alcohol, ether, wood spirit, the essential and fatty oils; it easily dissolves camphor, wax, fatty matters, India rubber, gutta percha, and a great number of resins. Amongst the last those which are the least soluble in it are shellac, copal, and animi. It is very inflammable, and burns with a smoky flame. Hydrogen gas passed through it, and charged with its vapor, burns with a very clear, luminous flame.

Chlorine and bromine convert benzole into the terchloride and terbromide of benzole. To the direct solar light, the change takes place very quickly. Concentrated sulphuric acid dissolves benzole, and when the mixture is gently heated a copulated acid, *sulpho-benzolic acid*, is formed, C^{12}, H^6, S^2, O^6, the hydrogen of which may be replaced by metals. As this acid is soluble in water, in purifying rough benzole with sulphuric acid, it is necessary to avoid using an excess of the acid, and also heating the mixture. A solution of chromic acid does not act on benzole, and is therefore a good agent for the purification. Concentrated nitric acid converts benzole into nitrobenzole, to the manufacture of which we proceed.

Preparation of Nitro-Benzole.

The preparation of nitro-benzole is accomplished on a large scale, by allowing a fine stream of benzole, and another of the strongest nitric acid, to run together in a worm or long glass tube kept well cooled. The two liquids react on each other on coming in contact, heat is disengaged, and nitro-benzole is formed.

Commercial nitric acid, mixed with half its volume of sulphuric acid, may be substituted for the concentrated nitric acid.

The nitro-benzole collected at the end of the worm, is first washed with water, then with a solution of carbonate of soda, and afterwards once more with water.

Properties of Nitro and Bi-Nitro-Benzole— Nitro-Benzole.

Nitro-benzole is a yellowish liquid, which, at 59° F., has a specific gravity of 1.209. It boils at 415°, 4 F., and cools at 37°, 4; it crystallizes in needles. Having an odor closely resembling that of the bitter almond, it has been largely used in perfumery for scenting fancy soaps, for which purpose it has one advantage over the oil of bitter almonds—it is less affected by the action of alkalies. Almost insoluble in water, it is very soluble in alcohol, ether, and essential oils.

Concentrated sulphuric and nitric acids dissolve

it, but it is precipitated by the addition of water. It is decomposed by a continued boiling with sulphuric acid; and under the same circumstances with concentrated nitric acid, it forms bi-nitro-benzole. Neither the alkalies in strong aqueous solution, nor quick lime, act on nitro-benzole; but an alcoholic solution of the alkalies, acts energetically and forms azoxy-benzole (C^{24}, H^{10}, N^2, O^3). By the action of nitric acid on this last substance a number of other interesting bodies are produced, which it is not necessary to describe here.

Bi-Nitro-Benzole.

Bi-nitro-benzole is formed when nitro-benzole is added, drop by drop, to a mixture of equal parts of fuming nitric acid and sulphuric acid, as long as the liquids will mix. If such a mixture be boiled for a few minutes, it becomes, on cooling, a thick magma of bi-nitro-benzole, which is easily purified by repeated washings with water. A single crystallization from alcohol will furnish this body in long brilliant prisms which melt at a temperature above 212°, and crystallize again on cooling in a radiated mass.

Bi-nitro-benzole is very soluble in warm alcohol. When a plate of zinc, well cleaned, is placed in a cold alcoholic solution of bi-nitro-benzole, and hydrochloric acid is added by degrees, we observe that the disengagement of hydrogen, which at first takes place, soon ceases, and at the same time the liquid takes a crimson red tint.* The reaction being completed, the excess of zinc is removed and the liquor is saturated by an alkali, which precipitates the oxide of zinc colored in deep purple. The precipitate is collected on a filter and washed with alcohol.

By distilling the highly colored alcoholic washings, washing the residue with cold water, then re-dissolving it in alcohol and evaporating it afresh to dryness, the new matter is obtained perfectly pure. The authors have given it the name of *Nitrosophenyline*, $C^{12} H^6 N^2 O^2$. When obtained as above, it is a black shining substance; when heated, it fuses and decomposes directly; it is almost insoluble in water, but freely soluble in alcohol and acids. An alcoholic solution containing only 0.2 per cent. is so deeply colored that by reflected light the solution seems opaque and of an orange red.

Concentrated hydrochloric and diluted sulphuric and nitric acids form magnificent crimson red solutions with nitrosophenyline, which is precipitated from them again unchanged by alkalies.

Bi-nitro-benzole treated with an alcoholic solution of sulphide of ammonium, is at first converted into nitro-aniline.

$$C^{12} H^6 (NO)^4 N = C^{12} H^6 N^2 O^4,$$

* Church & Perkins. Quart. Journ. Chem. Soc., ix. p. 1.

that is to say, aniline, in which one equivalent of hydrogen is replaced by one of nitrous vapor. Nitro-aniline crystallizes in yellow needles, which stain the epidermis like picric acid.

Transformation of Nitro-Benzole into Aniline.

(a). *By means of Sulphide of Ammonium.*—An alcoholic solution of nitro-benzole, after having been saturated with ammoniacal gas, is treated with a current of sulphuretted hydrogen. The liquor now becomes of a deep dirty green color, and deposits a little sulphur. It is now left twenty-four hours, during which time crystals of sulphur are deposited, the odor of sulphuretted hydrogen disappears, and is replaced by a strong ammoniacal smell. If distilled now to recover the alcohol, a good deal of sulphur is deposited, and it is impossible to continue the distillation long, on account of the violent bumping which ensues. It is, therefore, allowed to cool, and the sulphur is removed. On distilling the liquor again, more sulphur is deposited, which must also be removed. The process must be continued, re-saturating the liquor with sulphuretted hydrogen if need be, until a heavy oily matter (aniline) deposits, which must be separated from the liquor and re-distilled by itself. The aniline is thus obtained nearly pure.

Instead of using an alcoholic solution of nitro-benzole, and treating it successively with ammonia and sulphuretted hydrogen, the alcoholic solution of sulphide of ammonium may be prepared beforehand, and the nitro-benzole poured into it. A part is dissolved immediately, and the remainder by dryness in the course of the operation. It is sometimes advantageous, instead of waiting until the aniline separates, to add hydrochloric acid to the liquor in the retort until it is slightly acid, and then to distil almost to dryness, by which means chloride of aniline is obtained. This is decomposed by an excess of caustic soda, and the aniline set at liberty, is distilled off.

To avoid any danger from the bumping, a tinned copper still must be used, which should be heated by steam under a high pressure; at first the temperature should not exceed 162° F., but after some time it could be raised to 212° or 230° F.

The ammoniacal alcohol condensed in the worm may be re-saturated with sulphuretted hydrogen, and used over again with a new quantity of nitrobenzole.

(b). *Reduction of Nitro-Benzole by Nascent Hydrogen.*—In preparing aniline by this process, the nitro-benzole and zinc are placed in a vessel, and diluted hydrochloric or sulphuric acid is added so as to produce the disengagement of a small quantity of hydrogen. By degrees the nitro-benzole disappears, and aniline is formed, which remains in solution in hydrochloric or sulphuric acid.

To isolate it, an excess of caustic soda is added and the mixture is distilled; the aniline passes over with the vapor of water.

Beauchamp first recommended the employment of acetic acid and iron filings. He places in a retort 1 lb. of nitro-benzole, 1½ lb. of iron filings, 1 lb. of concentrated acetic acid. The reaction takes place without the application of external heat, the mixture becoming hot by itself, and the vapor being condensed in a receiver which must be kept well cooled. The condensed products consist of aniline, acetate of aniline, and some unchanged nitro-benzole. These are allowed to cool, and are then returned to the retort and again distilled to dryness.

The distillate is now treated with fused caustic potash, and the aniline separates as an oily layer, which must be removed and distilled once more.

The residue of the mixture of iron filings, acetic acid and nitro-benzole, which remains in the retort after the distillation, still contains a considerable amount of aniline; to obtain this, the retort must be washed out with water acidulated with sulphuric or hydrochloric acid, and the solution filtered, and then evaporated to dryness.

The dry residue is then mixed with quick lime and placed in an iron or refractory ware retort, and distilled, and the aniline thus obtained must be rectified.

(c). *Reduction of Nitro-Benzole by Acetate of Iron.* —Acetate of iron reacts on nitro-benzole and converts it into aniline, while the sulphate, chloride and oxalate of iron, have no action on it. The reaction is represented thus,

$$\underset{\text{Nitro Benzole}}{C^{12}H^5AzO^4} + \underset{\text{Acetate of Iron,}}{12FeO + 2HO + \overline{A}} =$$
$$\underset{\text{Aniline}}{C^{12}H^7Az} + \underset{\text{Acetic Acid.}}{\overline{A}}$$

One part of nitro-benzole is placed in a retort with an aqueous solution of acetate of iron, the retort is then heated over a water bath for several hours, and then the contents are filtered, being diluted with water if they have become pasty.

The residue left on the filter, which is principally peroxide of iron, is washed with boiling water. The filtrate and washings are then distilled. The condensed products being water, acetic acid, and acetate of aniline. These may be again distilled with strong sulphuric acid, using 4-10 the weight of the nitro-benzole employed to recover the acetic acid, and form sulphate of aniline, and the latter may be decomposed by caustic potash and the aniline distilled off. This process has not been found advantageous, and has consequently been given up.

(d). *Reduction of Nitro-Benzole by Means of Arsenite of Potash or Soda.*—In this process digest nitro-benzole with a solution of arsenious acid in a strong lye of caustic soda or potash, or place the arsenical solution in a tubulated retort, heat it to the boiling point, and then allow the nitro-benzole to fall drop by drop in it. Under these circumstances, nitro-benzole is transformed into aniline, which distils over, and it is only necessary to saturate with an alcoholic solution of oxalic acid to obtain perfectly pure oxalate of aniline.

CHAPTER X.

ANILINE PURPLE—VIOLINE—ROSEINE—EMERALDINE—BLEU DE PARIS.

§ 1. *Aniline Purple.*

It has been known for many years that hypochlorites react on aniline and its salts, producing a purple-colored solution; in fact, hypochlorites are the distinguishing test for aniline; but nothing definite was known of this purple-colored solution, it being simply stated that aniline produced with hypochlorites a purple-colored liquid, but that this color was very fugitive. Many absurd statements have been made respecting the discovery of aniline purple. We will just briefly mention how it was discovered by Mr. Perkins.

In the early part of 1856, he commenced an investigation on the artificial formation of quinia. To obtain this basis, he proposes to act on toluidine with iodide of allyle, so as to form allylo toluidine, which has the formula:—

$$\left. \begin{array}{l} C^7, H^7, \\ C^3, H^5, \\ H, \end{array} \right\} N = C^{10}, H^{13}, N.$$

thinking it not improbable that by oxidizing this, he might obtain the desired result thus:—

$$2\ (\underbrace{C^{10}\ H^{13}\ N}_{\text{Allyle-toluidine.}}) + O^3 = \underbrace{C^{20}\ H^{24}\ N^2\ O^2}_{\text{Quinia.}} + H^2\ O.$$

For this purpose he mixed the neutral sulphate of allyle-toluidine with bichromate of potash; but instead of quinia he obtained a dirty reddish-brown precipitate. Nevertheless, being anxious to know more about this curious reaction, he proceeded to examine a more simple base under the same circumstances. For this purpose he selected aniline, and treated its sulphate with bichromate of potash. This mixture produced nothing but a very unpromising black precipitate, but on investigating this precipitate he found it to contain the substance which is now, we may say, a commercial necessity, namely, *aniline purple.*

The method adopted for the preparation of aniline purple is as follows: Solutions of equivalent proportions of sulphate of aniline and bichromate of potash are mixed, and allowed to stand till the reaction is complete. The resulting black precipitate is then thrown on a filter, and washed with water until free from sulphate of potash. It is then dried. This dry product is afterwards digested several times with coal-tar naphtha until all resinous matter is separated, and the naphtha ceases to be colored brown. After this it is repeatedly boiled with alcohol to

ANILINE PURPLE. 83

extract the coloring matter. This alcoholic solution, when distilled, leaves the coloring matter in the bottom of the retort as a beautiful bronze-colored substance.

The aniline purple prepared according to the process just described, although suitable for practical purposes, is not chemically pure. If required pure, it is best to boil it in a large quantity of water, then filter the resulting colored solution, and precipitate the coloring matter from it by means of an alkali. The precipitate thus obtained should be collected on a filter, washed with water until free from alkali, and dried. When dry it is to be dissolved in absolute alcohol, the resulting solution filtered, and then evaporated to dryness over the water-bath. Thus obtained, aniline purple appears as a brittle substance, having a beautiful bronze-colored surface; but if some of its alcoholic solution be evaporated on a glass plate, and viewed by transmitted light, it appears a beautiful bluish violet color. If considerable quantities of an alcoholic solution of the coloring matter, containing a little water, be evaporated to dryness, the surface of the coloring matter next to the evaporating dish when detached, often possesses a golden green appearance. Aniline purple is, with difficulty, soluble in cold water, although it imparts a deep purple color to that liquid. It is more soluble in hot water, but its hot aqueous solution when left to cool assumes the form of a purple jelly. It is very soluble in alcohol, though nearly insoluble in ether and hydrocarbons. Aniline dissolves it readily. In properties, it seems to be slightly basic, as it is more soluble in acidulated than in pure water. Alkalies and saline substances precipitate it from its aqueous solution, as a dark purplish-black powder. Bichloride of mercury precipitates it in a very finely divided state; a little of this precipitate, which appears to be a double compound of chloride of mercury and coloring matter, when suspended in water and viewed by transmitted light, appears of a blue or violet color. If a small quantity of hydrates of potash or soda be added to an alcoholic solution of the coloring matter, it causes it to assume a violet tint, but without effecting any change in the coloring matter itself. Ebullition with alcoholic potash does not decompose it. Aniline purple dissolves in concentrated sulphuric acid, forming a dirty green solution. This, when slightly diluted, assumes a beautiful blue color. Excess of water restores it to its original purple color. We have had a specimen of this coloring matter heated for an hour to 100° Centigrade with Nordhausen sulphuric acid, without suffering decomposition, being restored to its original color by means of water, and possessing precisely the same properties as it had before being subjected to this powerful agent. Hydrochloric acid acts upon it in the same manner as sulphuric acid. It is decomposed by chlorine, and also by fuming nitric acid. Bichloride of tin is without action upon it. Powerful reducing agents have a peculiar action upon this coloring matter, somewhat analogous to the action of reducing agents on indigo. An alcoholic solution of the coloring matter when mixed with a little protoxide of iron changes to a pale brown color. This solution also becomes purple when exposed to the action of the atmosphere. Sulphurous acid does not affect the color of this substance.

This coloring matter forms a remarkable compound with tannin. When an aqueous solution of the coloring matter is mixed with a solution of tannin, precipitation takes place; the precipitate thus formed, after having been well washed, no longer possesses the properties of the pure coloring matter. It is insoluble in water. Like the pure coloring matter, it dissolves in concentrated sulphuric acid, forming a dirty green liquid, but on adding an excess of water to that solution, the new compound is precipitated unchanged. This compound is rather duller in color than the pure coloring matter itself. Aniline purple, when agitated with a little moist binoxide of lead, is transformed into Roscine. Its coloring matter is remarkable for its intensity; a few grains will color a considerable quantity of spirit of wine.

§ 2. *Violine.*

This coloring matter, which is a product of the oxidation of aniline, was first obtained by Dr. David Price. He prepares it by heating an aqueous liquid, containing two equivalents of sulphuric acid and one equivalent of aniline, to the boiling point, and then adding one equivalent of binoxide of lead, boiling the mixture for some time and filtering it whilst hot. The filtrate, which is of a dark purple hue, is boiled with potash, to separate the excess of aniline, and also to precipitate the coloring matter. When all the free aniline is volatilized, the residue is thrown on a filter and slightly washed with water, and then dissolved in a dilute solution of tartaric acid. This solution, after filtration, is evaporated to a small bulk, re-filtered, and then precipitated by means of an alkali. Thus obtained, violine presents itself as a blackish purple powder, which, when dissolved in alcohol and evaporated to dryness, appears as a brittle, bronze-colored substance, similar to aniline purple, but possessing a more coppery colored reflection. It is more insoluble in water than the preceding coloring matter; it is very soluble in alcohol; insoluble in ether and hydrocarbons: these solutions possess a color somewhat similar to that of the field violet. Concentrated sulphuric acid dissolves it, forming a green solution, but excess of water restores it to its original color.

Like aniline purple, reducing agents deprive it of its color, which is restored by the action of the atmosphere. Tannin produces an insoluble compound with it. When agitated with a small quantity of binoxide of lead, it is converted into aniline purple, excess of this reagent changes it into roseine.

§ 3. *Roseine.*

This substance nearly always accompanies aniline purple, though in very small quantities. It was first noticed publicly by C. Greville Williams, and afterwards by Dr. David Price. Williams used manganates for its preparation, but Dr. David Price prepared it by means of binoxide of lead. His process is as follows: To a boiling solution of one equivalent of sulphate of aniline, two equivalents of binoxide of lead are added, and the mixture boiled for a short time. The rose-colored solution is then filtered, and the filtrate evaporated to small bulk, which causes a certain amount of resinous matter to be separated; this evaporated solution is then filtered, and the coloring matter precipitated by means of an alkali, it is then collected on a filter, slightly washed, and then dried. The coloring matter thus prepared, readily dissolves in alcohol, forming a fine crimson colored liquid, which when evaporated to dryness, leaves the coloring matter as a dark brittle substance, having a slightly metallic reflection. It is much more soluble in water than either aniline purple or violine, but like them it is insoluble in hydrocarbons, and is more soluble in acids than in neutral liquids. Concentrated sulphuric acid dissolves it, forming a green solution; excess of water restores it to its original color. It forms a compound with tannin; and is also decolorized, or nearly so, by powerful reducing agents.

The three coloring matters just mentioned, namely, aniline purple, violine and roseine, are evidently closely allied, for they have nearly the same properties. They are all formed under similar circumstances, namely, by the action of oxidizing agents in the presence of water; they are all slightly soluble in water, though as the shade of color becomes redder, so their solubility increases; alkalies precipitate them from their aqueous solutions; concentrated sulphuric acid dissolves them, forming green solutions which an excess of water restores to the original color of the coloring matters; powerful reducing agents deprive them of their color or nearly so, but it is again restored by the influence of oxygen; and lastly, tannin forms insoluble compounds with them all.

§ 4. *Emeraldine or Aniline Green.*

Most chemists, who have worked with aniline in the laboratory, must have noticed the peculiar green-colored substance which forms on the outside of the various kinds of chemical apparatus that have been standing in the vicinity of any quantity of this body. This product is *aniline green*. It has been known for several years; it may be formed by various processes. One consists in oxidizing aniline with chloric acid; this is effected by mixing an hydrochloric solution of aniline with chlorate of potash. It may also be obtained by oxidizing a salt of aniline by perchloride of iron. Obtained by either of these processes, it presents itself as a dull green precipitate, which when dried assumes an olive green color. It is insoluble in water, alcohol, ether and benzole. Sulphuric acid dissolves it, forming a dirty purple-colored solution, from which it is precipitated unchanged by water. With alkaline solutions, it changes to a deep color somewhat similar to indigo, but acids restore it to its original color. The color of aniline green is much enlivened by the presence of an excess of acid, but unfortunately as soon as this acid is removed, it passes back to its normal color.

§ 5. *Bleu de Paris.*

This is another coloring matter produced under circumstances similar to those which give Futschine. MM. Persoz, De Luynes, and Salvetat give the following account of its preparation and properties: "9 grains of bichloride of tin and 16 grains of aniline heated for thirty hours at a temperature of about 356° F., in a sealed tube, produce neither a red nor a violet, but a very pure and lively blue.* Mr. Perkins repeated the experiment twice, but he obtained only a dirty green color; but at last he obtained the blue as described by MM. Persoz, De Luynes, and Salvetat. This blue crystallizes from the alcoholic solution in the form of fine needles, having the aspect of ammoniacal sulphate of copper; soluble in water, alcohol, wood-spirit and acetic acid; insoluble in ether and bisulphide of carbon. With concentrated sulphuric acid it forms an amber-colored solution, which water converts into a magnificent blue liquid. Strong nitric acid decomposes it, chromic acid precipitates it from its aqueous solution without decomposition, chlorine destroys it, sulphurous

---

* When you break the tubes in which the reaction has been effected, you obtain a blackish matter which, exhausted by boiling water, colors it blue; the solution, treated by common salt, left to precipitate the coloring matter that you collect on a filter, whilst the liquor takes a green shade more or less dark. The blue precipitate is redissolved anew in water, and precipitated again by the chloride of sodium. This operation is repeated several times to separate completely the green coloring matter, at last precipitate by few drops of hydrochloric acid, collect the blue matter on a filter, wash first with water acidulated with hydrochloric acid, then with pure water, the washing is terminated when the water begins to pass blue.

To obtain it crystallized, dissolve it in boiling alcohol, which, by cooling, deposits it in form of fine needles.

acid does not decolorize it, sulphide of ammonium is without action upon it. It is precipitated from its aqueous solution by alkalies and saline compounds. Submitted to the action of heat, it melts and decomposes in giving violet vapors.

CHAPTER XI.

FUTSCHINE, OR MAGENTA.

THIS beautiful product, which is often improperly called Roseine, is a member of an entirely different series of compounds from the foregoing, being formed under very different circumstances, and possessing very different properties. This coloring matter was first observed by Natanson, in 1856, when studying the action of chloride of Ethylene on aniline, and afterwards, shortly before it was practically introduced into the arts, by Dr. Hoffmann, when preparing cyantrephenile-diamine by the action of bichloride of carbon on aniline. It was M. Verguin who first brought it forward as a dyeing agent, and who, we believe, taught manufacturers how to prepare it on a large scale. Futschine is invariably formed at a temperature ranging from 17° to 19° Centigrade. It is produced from aniline by the action of reducible chloronized, brominized, iodized or fluorized substances, as well as by weak oxidizing agents. The substances used for its preparation on the large scale are perchlorides of tin and of mercury,

FUTSCHINE, OR MAGENTA.

and the nitrate of mercury. It has also been prepared with bichloride of carbon.

Preparation of Futschine by the action of Bichloride of Tin on Aniline.—Aniline combines with bichloride of tin, evidently producing a double compound. This product is a white substance, and may be prepared by adding to aniline, bichloride of tin in the anhydrous state or dissolved in water. Anhydrous bichloride of tin combines with aniline with great energy to form this compound. To prepare Futschine from the double compound, it is necessary that it should be free from water, or nearly so; therefore anhydrous bichloride of tin is generally employed for its preparation. The process adopted is as follows: anhydrous bichloride of tin is slowly added to an excess of aniline, the mixture being constantly stirred, and the pasty mass thus formed gradually heated; as the temperature increases, it becomes quite liquid and also brown in color. As soon as the temperature nearly approaches the boiling point, the mixture rapidly changes to a black-looking liquid, which, when viewed in thin layers, presents a rich crimson color; this is kept at its boiling point some time, and then well boiled with a large quantity of water; by this means the principal part of the coloring matter is extracted, together with considerable quantities of tin in the form of a protocompound. The aqueous solution of the coloring matter and hydrochlorate of aniline is then boiled, so as to volatilize any free aniline it may contain, and then saturated with chloride of sodium. The chloride of sodium causes the coloring matter to separate as a semi-solid, pitchy substance of a golden green aspect, while the hydrochlorate of aniline remains in solution. The coloring matter thus obtained, may be further purified by digestion with benzole, which dissolves out a certain amount of resinous matter.

Preparation of Futschine by the Action of Nitrate of Mercury on Aniline.—When protonitrate of mercury is left in contact with aniline for some time, it forms a white pasty mass, but when carefully heated to 170° or 180° Centigrade, it reacts upon it, forming a brown liquid, which gradually changes till of a dark crimson color. At the same time the whole of the metal of the mercury salt collects at the bottom of the vessel the experiment is conducted in. This product, when separated from the metallic mercury and allowed to cool, becomes semi-solid, being filled with crystals of nitrate of aniline. To purify this product it is best to dissolve out the nitrate of aniline it contains, in a small quantity of cold water, and then to boil the remaining product several times with fresh quantities of water, until the principal of the coloring matter is extracted, and filter the resulting aqueous solution while hot. On cooling, the solution will deposit the coloring matter as a golden-green, tarry substance, from which benzole separates a small quantity of a brown impurity, leaving the coloring matter as a brittle solid.

We have briefly described the above processes, because they may, to some extent, be regarded as types of most of the methods employed for the production of this coloring matter; the first, representing its formation, by the action of reducible chlorides upon aniline, and the latter by the influence of weak oxidizing agents.

Futschine is undoubtedly an organic basis, and a more powerful one than is generally supposed. The products obtained from aniline by means of bichloride of tin, is hydrochlorate of Futschine, and that obtained by the oxidizing action of nitrate of mercury, is the nitrate of Futschine. Our reason for stating this is, that on examining the coloring matter obtained by chloride of tin, it is found to contain large quantities of combined hydrochloric acid, and when nitrate of mercury was used, considerable quantities of combined nitric acid, therefore we conclude that the former is the hydrochlorate and the latter the nitrate.

Futschine is separated from its salts by precipitation with a small quantity of ammonia. When freshly precipitated, Futschine is a red, bulky paste, which, when dry, contracts, forming a purplish red powder. It is difficultly soluble in water, but an excess either of hydrochloric or sulphuric acid dissolves it, forming a brownish yellow liquid, from which ammonia separates it unchanged. By this reaction it may be distinguished from Roseine, which dissolves in strong sulphuric acid producing a green liquid. Caustic alkalies or ammonia in excess partially precipitate Futschine from its salts, but at the same time dissolve a considerable quantity of it, forming nearly colorless liquids. Acetic acid added to these alkaline solutions, restores the color of the Futschine; and if the liquids are concentrated, the bases precipitate it as a red, flocculent substance. An alcoholic solution of Futschine, when evaporated to dryness, leaves the coloring matter as a brittle mass, having a beautiful golden-green metallic reflection. By transmitted light it has a red color. Futschine has been analyzed, and is represented by the formula, $C^{12}H^{12}N^2O$.

In the hydrochlorate, Mr. Bechamp found a quantity of hydrochloric acid corresponding with the formula $C^{12}H^{12}N^2O\,HCl$. He also examined the hydrochloro-platinate which is a purple precipitate; it has the formula $C^{12}H^{12}N^0OHPtCl_3$. The existence of oxygen in this basis is remarkable, because, in many instances, it is produced from agents which do not contain a trace of oxygen, as, for example, bichloride of tin and aniline. The only way to account for the presence of oxygen in the product analyzed, is as an hydrate, thus:—

$$C^{12}H^{12}N^2O = C^{12}H^{10}N^2 + H^2O$$

Futschine.   Anhydrous Futschine.   Water.

This is, perhaps, to some extent confirmed by an experiment made with iodaniline. Iodaniline, when heated, yields Futschine; this change can be expressed thus:—

$$2(C^6[H^6I]N) = C^{12}H^{10}N^2 + 2HI$$

Iodaniline..   Anhydrous Futschine.   Iodhydric Acid.

But supposing the Futschine examined by Mr. Bechamp to have been an hydrate, it is remarkable that its hydrochlorate, and, more particularly its hydrochloro-platinate should also be hydrates; but as our knowledge of this body is as yet but scanty, we must wait for the accumulation of facts before we can form any fixed opinion respecting its constitution. The compounds investigated by Mr. Bechamp appear to be uncrystallizable. Reducing agents decolorize Futschine, but the oxygen of the air renders it its color. Like aniline purple, Futschine is a very intense coloring matter; tannin precipitates both Futschine and its salts, forming difficultly soluble substances. Bichloride of mercury precipitates this substance and its salts, forming double compounds; when preparing Futschine by means of bichloride of tin, there are two coloring matters produced, one possessing an orange color, and the other a purple hue. Little is known of them.

CHAPTER XII.

COLORING MATTERS OBTAINED BY OTHER BASES FROM COAL TAR—NITROSO-PHENYLINE—DI-NITRO-ANILINE—NITRO-PHENYLINE—PICRIC ACID —ROSOLIC ACID—QUINOLINE.

THE bases toluidine, xylidine, and cumidine, yield coloring matters under the oxidizing agents, and also when submitted to the action of reducible chlorides, at high temperatures, analogous to those obtained from aniline under similar circumstances, but the results generally are not so good, the color of the products becoming tinged with brown, as the bases get higher in the series.

Nitroso-Phenyline.

This remarkable body is obtained by the action of nascent hydrogen on an alcoholic solution of di-nitro-benzole. It is represented by the formula $C^6H^6N^2O$. This body is almost insoluble in water, but soluble in acids and in alcohol, producing crimson-colored solutions, but its color is not nearly so brilliant as that of Futschine. Any experiments with it, as regards its dyeing properties, have not been tried.

Di-nitro-Aniline.

Di-nitro-aniline is obtained by decomposing di-nitro-phenyle citra-conamide by means of carbonate of soda. When pure, it crystallizes in yellow tables. It dissolves very sparingly in water, producing a yellow liquid. It has the formula $C^6H^3(NO^2)N^2$. It does not combine with acids or alkalies, although it appears to be more soluble in acidulated than in pure water. Silk can be dyed yellow with di-nitro-aniline.

Nitro-phenylene diamine, or Nitro-azo-phenylamine.

Di-nitro-aniline, when submitted to the action of sulphide of ammonium, changes into this beautiful base, which crystallizes in needles of a red color, somewhat similar in appearance to chromic acid. It dissolves in water, forming a yellow or orange-colored solution like that of bichromate of potash. Alcohol and ether dissolve it freely. This base possesses the power of dyeing silk a very clear golden color.

Picric, or Dinitro-phenic Acid.

This beautiful acid was discovered as early as 1788, by Hausmann. It may be obtained by the action of heated nitric acid on a great variety of substances. The following are the names of some of them: Indigo, Aniline, Carbolic acid, Saligenine, Salicylious and Salicylic acids, Salicin, Phlorizin,

Cumanin, Silk, Aloes, and various Gum-resins. It is now prepared for commercial purposes from carbolic acid, and also from certain gum-resins. We have prepared it from carbolic acid on a large scale, in the following manner, with success: As strong nitric acid acts very violently, when brought in contact with carbolic acid, we have found it best to use an acid having a gravity less than 1.3, so as partially to convert the carbolic acid, and afterwards to boil it in stronger acid to change it into picric acid. On diluting the acid solution, the impure picric acid precipitates; to further purify this, it should be crystallized from boiling water. When preparing this product for commercial purposes, it is advantageous to let all the nitrous fumes formed in its preparation, together with a certain amount of atmospheric air, to pass over a fresh quantity of carbolic acid. This will absorb them and at the same time be converted into nitro, or di-nitro-phenic acid, and consequently diminish the quantity of nitric acid required for its manufacture.

When preparing picric acid from carbolic acid, there is always a quantity of a yellow, resinous matter produced, and at times a considerable quantity of oxalic acid. The latter is always produced when the acid which is used to finally convert the carbolic acid is too weak, for then it rapidly decomposes the picric acid, yielding carbonic and oxalic acids. Picric acid, when pure and dry, is of a light primrose-yellow color, crystallizing in strongly-shining lamina. It possesses an extremely bitter taste, and dissolves in water with a beautiful yellow color. When digested with protoxide of iron, in the cold, it yields a brown amorphous compound, which dissolves in water with a blood red color. Picric acid was introduced as a dye about five or six years since, by MM. Guinon, Marnas, and Bonney, eminent silk dyers of Lyons. Many of the cheap products sold as picric acid are of a brown color, and consist of impure di- and tri-nitro-phenic acids, and sometimes of this crude product and ground turmeric.

Rosolic acid.—Runge first noticed this substance in 1834, when studying creosote, but it was almost lost sight of, until again observed by Dr. Hugo Miller only a short time since. He accidentally observed that when crude phenate of lime is exposed to a moist, heated atmosphere, as that of an ordinary drying stove, it gradually changes in color, and assumes a dark red tint; this coloration is owing to the formation of rosolate of lime. Dr. Muller prepared rosolic acid from this product in the following manner: The crude rosolate of lime is first boiled with a solution of carbonate of ammonia. By this means a crimson solution containing the rosolic acid is obtained; this solution is then evaporated nearly to dryness, during such process ammonia is given off, and the crimson-colored liquid gradually changes to a yellowish red, and at the same time a dark resinous matter separates; the resinous substance is crude rosolic acid. In order to purify it, it is submitted to the following treatment, proposed by Runge: The crude rosolic acid is dissolved in alcohol, and by hydrate of lime in slight excess. The beautiful crimson solution which is thus formed is agitated for some time with the undissolved portion of the lime, filtered, and the filtrate diluted with water, and, lastly, the alcohol distilled off. The residuary rosolate of lime is then decomposed with just a sufficient quantity of acetic acid, and the whole boiled until every trace of free acetic acid and still adhering alcohol is volatilized. The rosolic acid separates first as a red precipitate, but when heated, cakes together, forming a dark, brittle substance, having a greenish metallic lustre.

It may be still further purified by solution in alcohol, to which a little hydrochloric acid has been added, and precipitation with water. Pure rosolic acid is a dark amorphous substance, possessing the greenish metallic lustre of cantharides. Its powder is of a red, or rather scarlet shade, which, if rubbed with a hard, smooth body, assumes a bright gold-like lustre. In thin layers, rosolic acid presents an orange color, when viewed with transmitted light, but with reflected light, a golden metallic appearance. When thrown down from an alcoholic solution with water, it forms a flocculent precipitate of a bright red color, resembling the basic chromate of lead. Concentrated acids, as acetic, hydrochloric, and sulphuric, dissolve rosolic acid, forming a brownish yellow solution, of which water precipitates rosolic acid unchanged. To cold water, it imparts a bright yellow color, and is more soluble in hot than cold water. Alcohol and ether dissolve it. With ammonia, caustic alkalies and caustic earths, it forms dark red compounds. These compounds are very unstable. No precipitates are formed with aqueous solutions of the rosolates, with the basic acetate of lead, or with any other metallic salt. According to Dr. Muller, it is represented by the formula $C^{23}H^{22}O^4$. Rosolic acid has been prepared lately on a large scale for the purpose of printing muslin. It was rosolate of magnesia which was employed. It is not used since the discovery of Futschine.

CHAPTER XIII.

NAPHTHALINE COLORS — CHLOROXYNAPHTHALIC AND PERCHLOROXYNAPHTHALIC ACIDS—CARMINAPHTHA — NINAPHTHALAMINE —. NITROSONAPHTHALINE — NAPHTHAMEIN — TAR RED — AZULINE.

THE beautiful hydro-carbon naphthaline, which has yielded such a long category of substances to the chemist, up to the present time has yielded nothing of practical importance to the dyer. From it, the following color derivatives having been obtained, namely: Chloroxynaphthalic acid, Perchloroxynaphthalic acid, Carminaphtha, Ninaphthalamine, Nitrosonaphthaline and Naphthamein.

Chloroxynaphthalic and Perchloroxynaphthalic Acids.

These acids were discovered by Laurent. They are produced by digesting the chlorides, namely: the chloride of chloroxynaphthyle and the chloride of perchloroxynaphthyle with an alcoholic solution of hydrate of potash. They are difficult to obtain in quantity. Mr. Perkins has not obtained satisfactory results in their preparation. They have the formula $C^{10}(H^5 Cl) O^3$ and $C^{10}(H Cl^5)$

O^3 respectively. They are regarded with great interest, as being very closely allied with alizarine, the coloring matter of madder; in fact they are viewed as chlor-alizaric acid. The synopsis is based upon the idea of alizarine having the formula $C^{10} H^6 O^3$, but it happens very unfortunately for this theory, that the formula of alizarine itself is still a disputed point. Chloroxynaphthalic acid is of a yellow color, insoluble in water and with difficulty soluble in alcohol and ether; it dissolves in concentrated sulphuric acid. This acid is a very sensible test for alkalies, being changed to an orange red by them. This may be shown by moistening paper with a weak alcoholic solution of this acid, drying it, and then exposing it to ammoniacal vapors. This will cause it to assume a red color.

The chloroxynaphthalates are described as possessing great beauty, and are of yellow, orange, or crimson colors. The potash salt is of a red crimson color, and slightly soluble in water; the baryta salt crystallizes in silky needles, having a golden reflection. The strontia, lime, alumina, and lead salts are of an orange color; the cadmium salt is a vermilion colored precipitate; the copper and cobalt salts are crimson; and the mercury salt is of a red brown color. Once some silk was dyed with a small quantity of chloroxynaphthalate of ammonia, which Mr. Perkins prepared, and found it to produce a good golden yellow color, of great stabilty under the influence of light. Perchloroxynaphthalic acid is a yellow, crystalline body, insoluble in water, but soluble in alcohol and ether. With potash or ammonia it forms insoluble salts of red or crimson color of great beauty.

Carminaphtha.

This coloring matter was also discovered by Laurent. It is obtained by heating naphthaline with a solution of bichromate of potash, and then adding sulphuric or hydrochloric acids. It is described as a fine red substance, soluble in alkalies, but precipitated from its alkaline solutions by means of acids. Mr. Perkins never obtained this product when oxidizing naphthaline.

Ninaphthalamine.

Ninaphthalamine is a name which has been given to a remarkable base which was noticed by Laurent and Zinin; but nothing was known of its nature until resubjected to investigation by Mr. Wood, who has both described and analyzed it and some salts. Its formula is $C^{10}(H^8 NO) N$, or naphthalamine in which H is replaced by NO. Mr. Wood prepares this base in the following manner: Sulphuretted hydrogen is to be passed through a boiling solution of dinitronaphthaline in weak alcoholic ammonia, until nearly all the alcohol has distilled off, which operation should occupy two or three hours. The residue is then to be boiled with dilute sulphuric acid, and filtered. The filtrate, on cooling, deposits an impure sulphate of ninaphthalamine in the form of brownish crystals which are purified by recrystallization in water two or three times. Mr. Perkins has found when crystallizing this salt, that it is best to use water acidulated with sulphuric acid. When pure, this sulphate has to be decomposed with ammonia, and the resulting precipitate of ninaphthalamine washed with water. Thus obtained, ninaphthalamine appears as a bright red-colored crystalline precipitate, which, when viewed under a lens appears as beautiful needles. It is very soluble in alcohol, producing a solution which, when diluted, is of an orange color slightly tinged with brown, not nearly so pure in color as that of nitrophenylinediamine. It is slightly soluble in water, and possesses the power of dyeing silk with a color somewhat similar to that of ordinary annoto. With acids it produces colorless salts. Its formula is the same as that of nitroso-naphthaline, though it possesses very different properties. As a dyeing agent we do not think it would be of any value even if it could be obtained cheaply.

Nitroso-naphthaline.

This peculiar body is a product of the action of nitrous acid on naphthalamine. It is prepared by mixing a solution of hydrochlorate of naphthalamine with nitrate of potash. From this mixture it separates a reddish brown precipitate. This, when washed with water on a filter and then dried, is dissolved in alcohol, filtered, and evaporated to dryness on the water-bath. Thus prepared, it is a crystalline, dark-colored substance, having a greenish metallic reflection. It is soluble in alcohol, and also in benzole, forming orange red solutions. When acids are added to an alcoholic solution of nitroso-naphthaline it immediately assumes a most beautiful violet color, as fine as aniline purple. Alkalies restore it to its original color. Silk may be dyed a beautiful purple shade with this substance, provided a certain quantity of hydrochloric or sulphuric acids be present. But what is most unfortunate is, that when the silk thus dyed is rinsed in water, the color immediately passes back to that of the pure nitroso-naphthaline, and also that the amount of acid required to keep up the purple shade if left in the silk rots it in a few days. Could this purple be fixed, nitroso-naphthaline would be a cheap and most useful dye. Mr. Perkins has endeavored to produce the sulpho-acid of nitroso-naphthaline, thinking that if such a compound could be obtained, it would possess a purple color, because it would be an acid itself. But although sulphuric acid does dissolve it, forming a blue solution, yet no combination takes place. He also endeavored to produce this desired result by treating sulpho-naphthalamic acid with nitrous acid, but obtained only nitroso-naphthaline, the acid of the sulpho-naphthalmic acid having apparently separated.

Naphthamein.

Piria observed that naphthalamine and its salts produced blue precipitates, afterwards becoming purple, when brought in contact with perchloride of iron, terchloride of gold, nitrate of silver, and other oxidizing agents. This product of oxidation he terms naphthamein. It is prepared by adding a solution of perchloride of iron to a solution of hydrochlorate of naphthamein. This mixture gradually changes and becomes blue, and after the lapse of a short time deposits a blue precipitate. This, when separated by means of a filter, is washed with water, which causes it to change in color, until a reddish brown purple. The filtrate from this substance contains proto-chloride of iron, and, according to Piria, chloride of ammonium. Naphthamein, when heated, fuses and decomposes, leaving a residue of charcoal behind. It is insoluble in water, sparingly soluble in alcohol, but more soluble in ether. It forms a blue solution with concentrated sulphuric acid, and is precipitated from this solution by means of water. Silk and cotton may be dyed with it, but the color of this compound is so inferior, as to render it useless as a dyeing agent.

Tar Red.

This coloring matter was discovered by Mr. Clift, of Manchester, in 1853. It is obtained by exposing a mixture of the more volatile parts of the basic oils of coal-tar and hypochlorite of lime to the air for about three weeks. Of the pure coloring matter we know nothing, except that with tannin it forms an insoluble, or difficultly soluble substance. With different mordants it yields different colors. It seems probable that this coloring matter is derived from pyrhole.

Azuline.

This substance, which is a beautiful blue dye, has been introduced within the last year. It was discovered by MM. Guinon, Marnas and Bonncy, of Lyons, who keep the process for its preparation a secret. It is obtained from coal-tar, but from which of its numerous derivatives is not known. This coloring matter is a brittle, uncrystallizable body, possessing a coppery, metallic reflection. It is very difficultly soluble in water, but soluble in alcohol, producing a magnificent blue solution, having but a slight tinge of red. With concentrated sulphuric acid it forms a blood-red liquid which, when poured into an excess of water, precipitates the coloring matter unchanged. Dilute acids have no effect upon azuline. Its alcoholic solution, when mixed with an alcoholic solution of hydrate of potash, also changes to a dull red color. This, when diluted with water, forms a purple liquid which is gradually restored to its original blue color by hydrochloric acid. With excess of ammonia, the solutions of azuline change to a reddish purple color. This ammoniacal solution, when treated with sulphide of ammonium, gradually assumes a dull, yellowish brown color. Iodine destroys the color of azuline. In color it is not quite so fine as chinöline blue, though far superior to Prussian blue.

CHAPTER XIV.

APPLICATION OF COAL-TAR COLORS TO THE ART OF DYEING AND CALICO PRINTING.

WE cannot enter fully into this subject, because we do not feel sufficiently acquainted with the various operations of the dye house or print works to do so, and also because the technical details of dyeing and printing operations would not, we think, interest the reader. We, therefore, propose to speak of the different processes employed for dyeing and printing with coal-tar colors, in general terms only.

Dyeing Silk and Wool.

Silk and wool can be dyed with all the coal tar colors, with the exception of the rosolates, these fibres possessing in most cases a remarkable affinity, if we may so speak, for these coloring matters. Many of them, as aniline purple, and violine, are taken from their aqueous solutions so perfectly by these substances that the water in which they have been dissolved is left colorless; in fact, silk and wool take them up so rapidly that one of the great difficulties the dyer has to contend with, is to get the fibres dyed evenly.

To Dye Silk with Aniline Purple, Violine and Roseine.

One process is applicable for dyeing silk with either of these coloring matters, and it is a very simple one. An alcoholic solution of the coloring matter required, is to be mixed with about eight times its bulk of hot water previously acidulated with tartaric acid, and then poured into the dye-bath, which consists of cold water slightly acidulated. After being well mixed, the silk is to be worked in it, until of the required shade. If a bluer shade than that of the coloring matter is required, a little solution of sulpho-indigotic acid may be added to the dye bath, or the silk may previously be dyed blue with Prussian blue, or any other blue, and then worked in the dye-bath.

To Dye Silk with Futschine, Picric Acid, Chinoline Blue and Violet.

This process is still more simple than the above, as it is simply necessary to work the silk in cold, aqueous solutions of these coloring matters. With futschine or picric acid, a little acetic acid may be used, but with chinoline colors, acids must be avoided. With picric acid, a very clear green color may be obtained by adding a little sulpho-indigotic acid to the dye-bath. We may mention that violine is not of such a fine color as that produced by aniline purple and indigo blue; and also that roseine is not such a good color as futschine, or magenta.

To Dye Silk with Azuline.

The dyeing of silk with this coloring matter is far more difficult than with the preceding, requiring to go through two or three different processes. The difficulty, we believe, arises from the insolubility of azuline in water. The process generally employed is to work the silk in a solution of the coloring matter acidulated with sulphuric acid, and when of a sufficient depth, to raise the temperature of the dye bath to the boiling point, and work the silk in it again. After this, the silk is well rinsed in water until free from acid, and worked in a bath of soap lather; it is then again rinsed and finished in a dilute acid bath.

To Dye Wool with Aniline Purple, Violine, Roseine, Futschine, etc.

This operation is generally conducted at a temperature of 5° or 6° Centigrade, and the dye-bath is composed of nothing but a dilute aqueous solution of the coloring matter required. Acids should be avoided, or only a very small quantity used, as the resulting colors are not so fine when they are employed.

Method of Dyeing Cotton with Colors of Coal Tar.

When aniline purple was first introduced, considerable difficulty was experienced in dyeing cotton so as to obtain a color that would resist the action of soap. Aniline purple is absorbed by vegetable fibres to a certain extent, and very beautiful colors may be obtained by simply working cotton in its aqueous solution; but when thus dyed the colors will not stand the action of soap. We have tried the use of tin and other mordants, but without any satisfactory result.

In 1857, Mr. Puller, of Perth, and Perkins, simultaneously discovered a process by which this coloring matter could be fixed upon vegetable fibres, so as to resist the action of soap. This process is based upon the formation of an insoluble compound of the coloring matter with tannin and metallic base in the fibre. To effect this the cotton has to be soaked in a decoction of sumach, galls, or any other substance rich in tannin, for an hour or two, and then passed into a weak solution of stannate of soda, and worked in it for about an hour. It is then wrung out, turned in a dilute acid liquor, and then rinsed in water. Cotton thus prepared is of a pale yellow color, and has a remakable power of combining with aniline purple.

The above process may be modified, for example: the stannate of soda may be applied to the cotton before the tannin, and alum may be used in the place of stannate of soda. To dye this prepared cotton with aniline purple it is only necessary to work it in an acidulated solution of the coloring matter; and when thus prepared the cotton will absorb all the coloring matter of the dye-bath, leaving the water perfectly colorless. It has been found that cotton thus prepared can be dyed with any coloring matter that forms insoluble compounds with tannin, therefore it is used for dyeing with roseine, violine, futschine, and chinoline colors.

Cotton may also be dyed a very good and fast color by mordanting it with a basic lead salt and then working it in hot solution of soap to which aniline purple has been added. Oiled cotton, such as is used for dyeing with madder, is also used in dyeing these colors. Cotton simply oiled, and before mordanted with alum and galls, also combines rapidly with these coloring matters; but as the color of the prepared cotton is generally rather yellow, it interferes sometimes with the beauty of the result. Cotton is sometimes coated with albumen, which is coagulated by the action of steam, and the albumen which covers the cotton dyed in the usual manner. We may mention that violine, roseine, futschine, and also the chinoline colors combine with unmordanted vegetable fibres, as well as aniline purple. Picric and rosolic acids are not applicable for dyeing cotton.

Printing Calico with Coal Tar Colors.

The process generally employed for printing with these coloring matters is simply to mix the coloring matters with albumen or lacterine, print the mixture on the fibre, and then to coagulate the albumen or lacterine by the agency of steam. Mr. Perkins and Mr. Gray, of the Dalmonach

Print Works, discovered the first process of applying these substances to fabrics in a different manner from the above. It consisted in forming a basic carbonate or an oxide of lead on those parts of the cloth which were to be colored, and then working the cloth thus prepared in a hot lather containing the coloring matter. Where the cloth was mordanted with the lead compound coloring matter was absorbed; but when unmordanted it was left white, because pure cotton is not dyed with these coloring matters in the presence of soap. This procss was intended for the application of aniline purple, for at the period of this discovery, the other coal tar colors were unknown. Colors, dyed by this process were very pure, but it had many disadvantages, which have caused it to be disused. Lately the process previously described for dyeing colors upon cotton prepared with tannin has been applied to calico printing. It consists in printing tannin in the fabric previously prepared with stannate of soda, and then dyeing it in a hot dilute acid solution of the coloring matter. By this means the parts of the fabric which are covered with tannin are dyed a deep color, but the other parts are only slightly colored. These are cleared by means of well known processes. These methods of applying these coloring matters is also modified by printing a compound of the coloring matter required and tannin on the prepared cloth, instead of tannin only, and then steaming the goods.

Method of Applying Aniline Green to Fabrics.

This process is interesting as being the first example of the production of coal-tal colors on the fabric itself.

The process is very simple. The design is to be printed on the cloth with a thickened solution of chlorate of potash, dried, passed through a solution of an aniline salt, again dried, and allowed to hang in a damp atmosphere. In the course of two or three days, the color will be fully developed. The color thus produced may be changed into a dark blue by the agency of soap or an alkaline liquid. The quantity of aniline used in this process is very small.

Application of Nitroso-naphthaline.

If cloth is printed with a thickened solution of a salt of naphthalamine, dried, and then passed through a solution of nitrate of potash, nitroso-naphthaline will rapidly make its appearance as a reddish orange color, but unfortunately the color thus obtained will not resist well the action of soap.

Of the numerous coloring matters of which we have briefly spoken, there are only few that are at present employed by the dyer and printer, namely; Aniline purple, Futschine, Picric acid and Azuline, but we think it probable that others of them will soon be introduced, such as the Bleu de Paris; and Nitro-phenylenediamine might be used for silk dyeing, as its color is good and it stand the action of light well. Unfortunately the chinoline colors though very beautiful are most fugitive. There has been an endeavor to introduce the chinoline blue of late, but although a considerable quantity of silk was dyed with it at first, it is now scarcely used, because when exposed to the sun for two or three hours the dyed silk becomes bleached. Aniline purple resists the light best, futschine and alpha aniline purple soon fade, especially on cotton. Aniline and bleu de Paris are not easily acted upon by light when on silk.

When the coloring matters of coal tar were first discovered, there was a great fear that the workmen engaged in their manufacture would suffer in health. All we can say is, that during the few years Mr. Perkins had to do with this branch of manufacture, there has not been a single case of illness among the workmen, that has been produced by any operation carried on for the production of aniline purple.

CHAPTER XV.

ACTION OF LIGHT ON COLORING MATTERS FROM COAL TAR.

WE think it will interest the reader to give him an extract of a paper published by our celebrated master, M. Chevreul, on this subject. We translate it literally from the *Comptes Rendus* of the Académie des Sciences, Seance of the 16th July, 1860, vol. li.

Two coloring matters recently produced are of frequent use, one to dye violet, and the other red violet. Both are obtained from aniline. This basis, under the influence of hypochlorites, gives the *violet*, and treated by the anhydrous bichloride of tin gives the *red violet*, or *futschine*. Any coloring matter cannot be compared to the Futschine for the brightness, intensity, and purity of the color. It dyes the silk in 1*st red violet*, *red violet*, 5*th violet*, and you can raise a gam from the white till the 11th shade, from the shade 4th till the 8th, we have the color called *rose*. Carthamine applied on silk gives, generally, colors from the 3*d red violet* to the red, it can be then two, three, four or five gams of my chromatic circle comprised between the color of the Futschine and that of the carthamine, both applied on silk. Before the futschine, carthamine was used to give the finest rose, but it was a rose less violet, whilst futschine gives a rose to the 5th violet of the red violet, or the 1st red violet, ordinary color of the rose.

The roses of cochineal are, for the brightness and intensity, to the roses of carthamine that these are to the roses of futschine. Ladies who like the rose must avoid to place themselves near those who wear the rose of futschine or cochineal, if they wear themselves the rose of carthamine. If thanks are due to the author of the discovery of futschine, it is not a reason to have this color applied on silk used for curtains, tapestry, etc., for if futschine *has the beauty of the rose it has also its fragility.* It is enough of four hours of exposition to the sun, to have the silk dyed with futschine to become tarnish, turn vinous, and afterwards reddish.

Futschine on cotton is not stable. A card of specimens of wool, silk, cotton, dyed with futschine and carthamine, shows that *futschine* applied on silks is inferior in stability to the *carthamine*, for the silk dyed with this latter has an orange color more sensible that the one dyed with futschine, which has a violaceous color, and, however, that one had been raised to the 8th shade, whilst the specimen dyed with carthamine had been only to the

6, 5th shade. When the red violet of futschine is changed after four hours exposition to the sun, the red violet of cochineal has not changed after one week to the same exposition. Silk mordanted with alum and cream tartar and dyed in red violet, 9th shade, that is the shade above crimson, after an insolation of eight months has lost only 3 shades. At last silk dyed in 1st red violet, 10th shade, with cream tartar and tin composition lost in the same length of time 1-5th shade.

I have demonstrated in 1837 the influence of oxygen atmospheric in about every case, which, in stuffs dyed with organic coloring matters, are discolorized by their exposition to the sun, in proving that the same can be kept several years in luminous vacuo. I have demonstrated, in the same year, that, on the contrary, Prussian blue is decolorized in luminous vacuo; it becomes first white, then brownish, and is recolorized by the contact of oxygen. To-day I present to the Academy results very different; they have been given by *picric acid* used in dyeing since about 20 years.

Cold it gives to the wool, yellow, 8th shade; to the silk 2d yellow 5th shade. Boiling it gives to the wool the 3d orange yellow 9th shade, to the silk the 1st yellow 6th shade; in both cases it does not fix to the cotton. It is very curious to follow the changes that the wool and silk experience under the influence of luminous air; they are described in the following table:—

Color of the Silk.

After 6 days' insolation	yellow		9th shade.
" 18 " "	5th or. yellow		9th "
" 1 month "	4th "		9-5th "
" 2 " "	3d "		9th "
" 3 " "	3d "		9-8th "
" 4 " "	1st "		7-5th "
" 5 " "	1st "		7-5th "
" 6 " "	"	1-10th 6-25th	"
" 8 " "	5th "	2-10th 3d	"

Color of the Wool.

After 6 days' insolation	3d orange yellow	9-5th shade.
" 18 " "	3d "	9-5th "
" 1 months' "	2d "	10th "
" 2 " "	orange yellow	10-5th "
" 3 " "	"	" "
" 4 " "	5th orange	11th "
" 5 " "	4th "	10-75th "
" 6 " "	3d "	10-75th "
" 8 " "	3d "	11th "

These results are curious when you compare them to the proceedings. This progression by which the wool in 8 months gained 2 shades in passing from the 5th orange yellow 9th shade, to the 3d orange 11th shade, that is, passing by 8 gams towards the red. The silk, after gaining 4 shades, almost near the red, has begun to descend from the 3d month.

Reflections.

This is an important question to know if in the trade the buyer is not exposed to pay very dear, a color beautiful without doubt, but having no stability whatever, in the quality of the tissue.

This inconvenience is a real one, and this reflections have for object not to destroy but attenuate them.

Industry is free to manufacture any kind of colors, except in the case of a special convention between the manufacturer and the buyer.

The merchant cannot be responsible, but it is to the buyer to have the merchant indicate on his bill the name of the matter used to dye the stuff, by example if it is a crimson or a rose that the buyer wants sold, he will have the bill with the denomination of *crimson* or *rose* of cochineal. I speak here only for stuffs used in tapestry, and I do not refer to the roses of futschine and carthamine employed for dresses.

If buyers were knowing the difference which exists between stuffs of the same color, but dyed with different matters, we are certain that before long, our stores will not have other colors than those known to be solids; and if in a public place, the public had on the eyes two comparative tables, one dye with all colors which have been exposed to the sun a certain length of time, and the other with the same colors kept in the dark, the public will be soon instructed of the extreme difference existing between colors, and this instruction will be the best warrant to not be deceived in the trade of colors. We hope to see some enterprising houses establish such tables, and we are sure they will render a great service to the public at large.

CHAPTER XVI.

LATEST IMPROVEMENTS IN THE ART OF DYEING. CHRYSAMMIC ACID — MOLYBDIC AND PICRIC ACIDS—EXTRACT OF MADDER.

Chrysammic Acid.

LATELY a color prepared with aloes has been used to dye, and its fine properties deserve to attract the attention of dyers. Messrs. Sacc and Schlumberger have given a great attention to this product. We shall give its preparation and its uses to dye as described by Schlumberger.

Preparation of the Coloring Matter.

In a retort of a capacity of 22 to 28 gallons, introduce 67 pounds of commercial nitric acid and add to it about 18 ounces of aloes of the best quality. Heat the retort in a water bath under a chimney, when nitrous vapors begin to disengage, take out the fire and introduce in the retort by small portions 10 lbs. of aloes. When all the aloes has been introduced and the disengagement of nitrous vapors has stopped, pour the whole in a flat dish and evaporate in paste in a sand bath, and terminate the evaporation to dryness in a water bath. Put the mass on a filter and wash it several times with cold water and dry at a gentle heat.

The product in dye is of about 66⅔ per cent. of the aloes used. The cost for 2¼ pounds are about $1.40.

Dyeing of Wool with Chrysammic Acid.

If you dissolve in a kettle full of river water, 2 lbs. 12 ounces of aloes purple, that you boil and refresh, and introduce in this bath 34 pounds of well washed wool, this wool, after an hour of ebullition, takes a fine brown color. If the quantity of chrysammic acid is double, you obtain a fine velvet black.

If you dissolve 1 pound 11 ounces of chrysammic acid in water, to which you add $2\frac{1}{10}$ lbs. of calcined soda, you obtain a liquid of a very fine purple color, which after a few days is very intense, and which can communicate to 34 pounds of wool, by an ebullition of half an hour, a fine bluish color. The wool wants to be well washed; but do not require any mordant. If for the same quantity of wool you use the double of purple of aloes, you obtain a blue similar to the blue of indigo by the vat.

If you neutralize the filtered liquor collected from the washings of chrysammic acid obtained by evaporation, with a paste of chalk, and you filter the neutralized liquor, you can obtain with this liquor, several shades more or less light of olive green, according to the concentration of the bath.

At last chrysammic acid receives again a very important application, in the use of it to fix other colors which are not solid.

If you add 6¾ lbs. of orseille and 9 ounces of purple of aloes dissolved in caustic soda, you obtain an orseille color on which air and light have no action.

The extract of orseille found in the trade, communicates to wool brighter colors than common orseille, but they are not solid. Mr. Schlumberger has found that in mixing 11¼ lbs. of this extract with 18 ounces of dry aloes purple, and leaving the mixture several days, the colors obtained are solid and kept all their beauty.

Chrysammic acid then is one of the most solid colors that the wool dyer can find, and it deserves a more attentive study.

Molybdic and Picric Acid.

1. It is only since a short time that molybdic acid is used in the art of dyeing and different modes for its preparation have been indicated.

The molybdic acid can be prepared in the following manner. Melt together equal weights of molybdate of lead reduced to fine powder, with calcined soda, in an iron crucible, decant the formed molybdate of soda, then prepare with hot water a concentrated solution of this molybdate that you decompose by an excess of nitric acid, and you boil till the molybdic acid separates in the form of a fine yellow precipitate; this precipitate is washed with water and at last dried.

The molybdate of ammonia is prepared in the following manner: Introduce little by little in caustic ammonia, molybdic acid, as much as it can be dissolved. The dissolution of molybdic acid is accompanied by a disengagement of heat, and presents itself in the form of a light yellow color, which has a very strong ammoniacal smell, and must be kept out of the contact of the air.

I give now the different processes to dye stuffs with these preparations.

Dyeing of Silk.

You can obtain a very dark blue in impregnating silk with molybdate of ammonia: you leave to dry, and pass in a bath of hydrochloric acid, and immediately, without washing, in a bath of chloride of tin, to develop the blue color; wash well and dry. You can obtain lighter shades in diluting the molybdate of ammonia with water. Silk impregnated with a solution of molybdate of soda, at 20° B., dried and pass in hydrochloric acid and chloride of tin baths, takes a nice blue color. In diluting the molybdate of soda with water, you can obtain lighter shades.

These colors are very solid to the light.

Dyeing of Cotton.

The color on cotton appears less fine than on silk. The finest and darkest blues are obtained with the molybdate of ammonia; but, if the bath is diluted with three times its volume of water, you have then a gray-blue.

We have not the least doubt that before many years this substance will be used by all the profession.

2. Picric acid has been employed first by Mr. Guinon of Lyons, France, in the dyeing of silk and wool. Its process, to manufacture it by treating coal tar by nitric and sulphuric acids, he obtains a resinoid matter, which, dissolved in more or less water, gives the shade wanted. It is in this bath, heated at 105°, that he passes the silk without mordant, and he introduces it afterwards in the warm room, without washing, to fix the color.

The process to prepare it consists in heating coal tar, and to introduce into it three times its weight of nitric acid; that you let run in it by a small glass pipe: boil with the acid till in a syrupy consistence; wash several times with cold water, and afterwards with warm water, to separate the acid from the resinoid matters, and evaporate it to dryness to obtain crystals.

$15\frac{1}{2}$ grs. of picric acid, dissolved in a sufficient quantity of water, could dye, in yellow, $2\frac{1}{4}$ pounds of silk.

Silk cloths take in it a very fine shade, without altering their brightness. The results are the same with wool. With potash the shades can vary till yellow orange.

For more details on this acid, we refer to Chapter VII.

Madder.

Madder is one of the coloring matters which has been the most studied in these last times. That plant has been submitted to many treatments in order to extract from it its pure coloring matter. We shall enumerate briefly some of the most important treatments which have been tried on this plant.

Extract of Madder by Messrs. Julian and Roguer.

They operate on madder in powder; they shake it conveniently in large vats, with cold or hot water, deprived of calcareous salts. They run it in *vat-filters*.

According to the colors they wish to obtain, they leave the madder thus in paste in the vat-filters from one to five days, according to the want or not of an alcoholic fermentation. This paste is then well pressed and carried into ovens to be dried. The water collected after the pressure is submitted to the alcoholic fermentation.

Extract of Madder by Koecklin.

His process gives an extract of madder free of ligneous matters, and the colors obtained in dyeing are as good and solid as madder itself. He uses the neutral organic oxides, such as acetone, hydrate of methylene, alone or combined with alcohols or heterogenous substances. These oxides are used as solvents of the coloring matter.

It is by maceration and expression that he saturates the solvent; the bath being saturated, he precipitates the coloring matters by water, *i. e.*, till water does not produce any precipitate. The precipitate filtered and dried constitutes the extract of madder.

It is a known fact, that in the use of madder in dyeing, they utilize only two-thirds of the coloring matter, the last is retained in the residuum. Mr. Schwarts tried many experiments, the object of which was to utilize this coloring matter, and he has not succeeded. The best process he found is the following:—

He takes 7 pounds 14 ounces of commercial sulphuric acid, and reduces it at 60° B.; after it is cooled, he adds to it $6\frac{1}{2}$ ounces of flour of madder, which is equivalent to 13 ounces of washed madder. He leaves to macerate half an hour and throws the whole on a flannel: the filtration is slow, and the filtrate is of a very dark orange color. He pours this liquid in half a gallon of water, which precipitates all the coloring matter, and then filters a second time through a thick flannel cloth. The filtrate is an acid which marks 35° B.

The two matters left on the filters are perfectly washed with water, dried and weighed, they give three ounces of residuum, with a tinctorial power equal to six ounces of madder, and half an ounce of extract, equal to fifteen ounces of madder. For the acid at 35°, it can be used again in bringing it at 60° by distillation.

CHAPTER XVII.

THEORY OF THE FIXATION OF COLORING MATTERS IN DYEING AND PRINTING.

THERE are two methods of coloring stuffs which must not be confounded with each other. By one of these, the coloring matters, lakes, etc., are mixed with gums or varnishes to make them into a color which is applied to the stuff, and which, on drying, adheres to it. Whether these coloring matters are mixed with a fat varnish, drying oil, white of egg, the result is always the same; but this operation, which is purely mechanical, and which may be performed on every kind of fabric, will only occupy the printer's attention so far as relates to the discovering of that glutinous body which is most capable of rendering this or that colored substance adherent to such or such fabric. By the other method the coloring matters, brought to the proper conditions, are deposited and then fixed on the goods in such a manner as to be incorporated with the fibre, and only to be capable of being detached from it by the intervention of a more or less powerful chemical agent; but some of them —and in this number are several substances of the organic kingdom, such as indigotin, carthamin, curcumin, and among the mineral colors, the oxides of iron, chromium, lead, etc.— only require to be applied on the goods; whilst a greater number of others, such as madder, cochineal, Brazil and Campeachy woods, quercitron, and weld, unite with the different fibres only by the co-application of auxiliaries, which are designated by the name of *mordants;* it is in consequence of this difference that all who have written on dyeing have divided coloring matters into *those which adhere to the goods of themselves*, and *those which can only be fixed by the co-application of mordants.*

To discover the cause in virtue of which the different colored bodies unite with the textile fibres of cotton, wool, and silk, to such a degree as to form with them one body; to explain how it happens that one and the same substance has not the same aptitude for each of these fibres—such is the question which first presented itself to the scientific men who devoted their attention to the application of colors, and the solution of which is more especially important to the art of dyeing, of which the *printing* of fabrics is but a particular case. HELLOT and LE PILEUR D'APLIGNY, MACQUER, BERTHOLLET, BERGMANN, and CHEVREUL, who are justly entitled to rank as high authorities on this subject, have given forth different opinions on this point. The first two saw in the fixation of the colors on the goods only a purely mechanical operation; the last four, on the contrary, only an operation purely chemical.

Of all chemists Mr. Chevreul is the one who has searched most deeply into this important matter, and in comparing the general phenomena of dyeing with those which natural philosophers and chemists generally consider as dependent on molecular forces, the causes of chemical action, he arrives at the conclusion, that the first are of the number of those which take place when two or more bodies are in contact and their combination is effected slowly.

It appears therefore that whilst HELLOT and D'APLIGNY attribute all the effects produced by coloring matter, to the existence in the fibres, of pores more or less numerous and spacious, in which the coloring matter lodges, all chemists repudiate this view, and trace the same effects to chemical affinity.

Such were the notions entertained by scientific men on the causes of the adherence of coloring matters to the goods, when the views of Mr. WALTER CRUM were published. According to the experiments of DE SAUSSURE, experiments so full of interest and so well known, chemists were aware that charcoal absorbs gases without altering their nature, in proportions which vary according to the nature of these gases, its own nature, and its state of porosity. No one is now ignorant of the applications which are daily made of this body in the arts, for decoloring syrups, by freeing them from different substances. It is in connection with this order of facts, and enlightened, moreover, by the theoretic works of the celebrated chemist of Berlin, that Mr. CRUM proceeds to adduce arguments in favor of the ideas of *Hellot*. He advances, in fact, after passing in review the different modes of action of porous bodies, that several dyeing operations depend on the capillary action described by de Saussure; and this opinion he bases chiefly on the result of the microscopic examination of the fibres of cotton, which was made by Mr. Thompson, of Clitheroe, and M. Bauer—this examination having established that these fibres are formed of transparent and glass-like tubes, which, though cylindrical before their maturity, flatten, on the contrary, from end to end, as they ripen, and then present the aspect of two separate tubes. Mr. Crum thinks that, since the sides of these tubes permit water to pass through, they must be porous; but he adds, that neither the form, nor even the existence of such lateral perforations have been capable of being discovered by the aid of the most powerful microscope. This, as will be seen, is the hypothesis put forward by Le Pileur d'Apligny, presented under a new form, and with the reserve of a mind essentially experimental. This being assumed, the eminent Scottish manufacturer explains the fixation of the colors in the following manner.

He first admits that the mineral base of a madder-dyed color—oxide of iron or aluminium—treated with a volatile acid—*acetic acid*, for example—gives rise to a solution which, when impressed on the fabric, is there gradually decomposed in course of time, abandoning its acid, *just as it would be decomposed in similar circumstances without the intervention of the cotton;* and if this base, deposited on the fabric, remains adhering to it so powerfully as to resist the action of the most perfect washing, it is because the solution, after having penetrated by the lateral openings into the interior of the tubes which compose the cotton, is there decomposed, and the oxide being set free in the narrow passage where it is enclosed, can no longer be disengaged from it. When the cotton, then, composed of *sacs* thus lined with metallic oxide, passes into a madder-bath, or one of any other coloring matter, the latter combines with the metallic oxide by a true chemical action to form a lake, or what is properly called a color.

Such are, in few words, the principal considerations which this chemist brings to bear on the question. Persoz holds a different opinion, and proceeds to examine how far this theory, which, by the author's admission, has several points of resemblance to that of Hellot and Le Pileur d'Apligny, admits of being supported by the facts on which it is based. The following are Persoz's views on the subject:—

According to the first proposition, the acetate of alumina, for example, would be decomposed in presence of the goods, just as if it were free, and experience seems to him to be here opposed to such an assertion. He does not dispute that this salt, free, or in presence of the goods, is composed of acetic acid and alumina, or basic acetate; but that, for equal quantities, and diffused over equal surfaces of cotton cloth, plates of glass, mica, or platinum, and dried, moreover, in the same conditions, this acetate gives up always the same quantity of alumina, is what he finds it impossible to admit. In fact, if the desiccation takes place at a temperature but little elevated, the quantity of the earth, taken from the acetate by the cotton, will be incomparably greater than that which would be liberated on the glass or mica plates; it must be concluded, therefore, that the textile fibre of the cotton exercises a powerful influence on the decomposition of the acetate of alumina. But if any doubt still exist as to the part which the fibre performs in the decomposition of a mordant, the subjoined fact ought, he thinks, to dispel them. A solution of cubical alum, submitted to spontaneous evaporation, yields crystals of *cubical alum;* but if one puts in it, for a certain time, stuffs of silk and cotton, this same solution now furnishes, after undergoing a spontaneous evaporation, nothing but octahedral crystals of alum, deprived as it is by these stuffs of a notable portion of its base.

The organic and inorganic kingdoms, especially the former, furnish a great number of substances which possess the property of dyeing stuffs, either constituting colors by themselves, or entering as elements into compounds of a more complicated nature; but, to receive an application, these substances, simple or complex, must unite, if not by themselves, at least by the intervention of a suitably selected body, two essential qualities: first, *that of being insoluble or nearly so;* second, *that of resisting as much as possible the destructive action of the air and the solar rays.* The first of these qualities is indispensable; for if it be wanting, there is *coloration* of the goods, but not *dyeing*, in the proper sense of the word; a simple washing with water suffices to discharge the color. The second is not essential in the same degree, since it is subordinate to the stability which is intended to be given to the colors applied to a fabric.

Indigotin, carthamin, curcumin, oxide of iron, oxide of chromium, sulphide of arsenic, sulphide of antimony, are dyeing substances by themselves. When one interrogates experiment as to the means of making them adhere to the goods, so strongly as to constitute one body with them, it is found to be necessary either to form these colors on the stuff itself, by putting in presence of the latter the elements of which they consist, and one of which at least must be soluble, or, if these tints are previously formed, to make them enter into a soluble combination with which one impregnates the fabric to set them afterwards at liberty, in such a condition that they combine with the fabric in the nascent state, either as protoxide, which, by oxidizing in the air, passes by degrees into the state of sesquioxide, or in the state of sesquioxide at first. The color of sesquioxide of chromium is fixed only in the same conditions. Again, to make the sulphides of antimony and arsenic adhere, it is sufficient to apply to the goods one of the saline and soluble combinations of these bodies, then to decompose it by an acid so as to set them at liberty. The fixation of carthamine takes place under circumstances nearly similar.

The greater part of coloring matters—nine-tenths at least—are not of a dyeing power by themselves, and only become so by entering into a combination which has for its object, not only to give them the first quality essential to every tint for being fixed, *insolubility*, but oftener also to make them contract a shade which they do not assume by themselves. The coloring matter of madder, for example, which is soluble in water, acquires the property of dyeing only in so far as it is combined with a body capable, in the first place, of forming with it an insoluble compound, as certain fatty substances, the oxides of aluminum, tin, iron, *et cetera*, and then making it contract the hue which one desires to obtain.

The different dye woods do not dye better by themselves than madder; and they require, like it, to enter previously into a combination.

Chromic acid itself, rich as it is in color, becomes a dyeing substance only so far as it forms part of a saline combination, which should present, along with the shade desired, the greatest possible insolubility. Even the alumina, which serves as a base to all the organic colors, is not capable of fixing the chromic acid.

It is only in so far as they are formed on the stuffs themselves, that the dyeing compounds of this group become adherent to them. In any other case there is no dyeing, unless, as sometimes happens, the combination becomes by slow degrees insoluble, either by itself—*carthamin*—or by the intervention of a suitable agent—*catechu*. Experience proves, moreover, that of the two substances which usually occur or co-operate to the formation of the color, it is that which is insoluble which should be fixed first on the fabric, and with the same precautions as if one were dealing with one of the substances which are of a dyeing nature when used by themselves. The dyer deviates from this rule, only in so far as the elements of the lake, happening to be equally soluble, and endued moreover with an equal inclination for the fibre of the stuff, render it a matter of indifference whether the latter be first impregnated with the one or the other: thus the colored combination which is formed by nut-gall and a ferruginous preparation, is rendered adherent either by first depositing the iron compound on the fabric, and afterwards passing the latter into a decoction of nut-gall, or by commencing with impregnating the stuff with this infusion, to pass it afterwards into a ferruginous preparation.

This rapid glance at the formation and fixation of dyeing substances, will doubtless suffice to make it understood that the subject under consideration presents different orders of facts, which it is necessary not to confound. In the fixation of indigo, for example, there are, on the one hand, the formation of indigo-blue, and on the other, the adherence of the latter to the stuff. The first of these facts enters into the phenomena of oxidation that are best defined; the second into those of adherence or juxtaposition, which are confounded more or less with the facts pertaining to the aggregation of similar particles. In the fixation of the color of madder, and of all its congeners, there are in like manner two orders of facts: the one which relates to the most clearly understood chemical actions—namely, the union of this coloring matter with the oxide, which is called in to give it, besides the insolubility necessary to it, the desired shade; the other, which consists in the juxtaposition and adherence to the stuff, of the lake which it produces. So, in the fixation of chromic acid, considered as a coloring matter, it is necessary to distinguish between the formation of the colored saline compound which one wishes to obtain, and its fixation, properly speaking, on the fabric. There are, therefore, in all the operations of dyeing and of the fixation of the colors, certain phenomena, which, inasmuch as they belong to the most common chemical reactions, cannot give rise to any discussion; let it now be considered whether it be not possible to dissipate likewise all uncertainty in what concerns the others.

CHAPTER XVIII.

PRINCIPLES OF THE ACTION OF THE MOST IMPORTANT MORDANTS.

HITHERTO, the term *mordant* has been applied to every substance which possesses the twofold property of uniting, on the one hand, with the goods, and on the other with the coloring matters. From this, it might appear that the mordants possess properties quite peculiar, whilst in reality it is not so. Placing one's self in the point of view which accords with the theory advanced by Persoz, one sees in these bodies only the elements, the constituent principles, of a saline compound which forms on the fabric itself to become adherent to it.

From the fact that the colorable and colored principles all combine with the metallic oxides to form insoluble compounds, it would seem also that these last should all be capable of fulfilling the part of mordants, and, consequently, of becoming the base of the colored lakes formed on the stuff. It is not so, however; the number of bodies which possess this property is very limited. They are, among the compounds of the inorganic kingdom, the oxides of aluminium, iron, chromium, and tin; among the products of the organic kingdom, the modified fatty bodies. The Editor has already pointed out a resemblance of the oxides of aluminum, iron, and chromium among themselves, observing that the volume of their equivalents is the same; considered under another relation, these three compounds are, of all the metallic oxides, those which exhibit in the highest degree the property of passing from a state in which they possess their full aptitude for combining, to an isomeric state in which they become indifferent in the presence of the most energetic agents.

For a body to be capable of performing the part of a mordant, it is necessary, in accordance with the views already stated, that the dimensions of its molecules be in a simple ratio to those of the surface of the fibre, and that, being fixed on the fabric, it give rise to a colored compound, the faces of which, being also in a simple relation with those of the fibre, cause its adherence.

All the mordants do not in the same manner render the colors adherent to the stuffs; some cause them to undergo only slight changes of shade, depending on the acid or basic character which the mordant performs, and especially on the dimensions of the colored molecule which is formed. Thus, let hydrate of lead, on the one hand, be deposited on a stuff, and on the other, hydrate of alumina, both colorless, but possessed of different properties, and let this stuff be passed into a bath of cochineal; the aluminous mordant will be dyed red, and the lead mordant a deep black. The same will be the case, and for the same reason, with hydrate of tin and hydrate of alumina, which, if fixed on a stuff and dyed in a madder bath, will give—the latter, a red inclining to rose-violet, the former, a red inclining to orange. The others, particularly the oxide of iron, cause the colorable or colored principle to previously undergo an alteration; for, if the iron oxide combined purely and simply with the coloring matter of the madder, for example, which in its state of isolation is of a clear brown or orange-yellow, one should obtain lakes of a clearer color than that which is peculiar to this oxide, whilst lakes are produced of which the shade varies from the most intense black to the most delicate lilac, according to the proportion of oxide on the stuff.

The nature of the principal mordants being known, the first point to be investigated is this—whether it be a matter of indifference to employ one saline combination rather than another, to render their base adherent to the goods? There are, in this question, two points to be considered: the first is one which the manufacturer should never lose sight of in the operations by which he applies a mordant on the goods, namely, the chemical part which this mordant, once fixed, ought to fulfil in presence of the coloring matter. Suppose, for example, that instead of having set at liberty on the goods hydrated alumina in that state in which it has all its chemical properties, it has, in point of fact, been deposited thereon in that state in which it loses momentarily all its aptitude for combining—the operation will be a failure, and goods thus mordanted will not dye. The second point is this, namely, that the brightness and intensity of the color which is obtained from a mordant depend on the manner in which this mordant is set at liberty, and passes into the insoluble state on the fibre, to be brought into immediate contact with it. Thus, let hydrate of alumina be prepared with every precaution, let one part of it be slowly dried, and another quickly, and there will be obtained, in the first case, a coherent mass of a horny aspect, in the second, a dull and opaque mass; and these two pieces, immersed in a solution of coloring matter of pure madder, will be dyed, the one of a red almost brown, the other, a dull and pale red. It is important, therefore, to seek, among saline combinations, that which yields most easily to the goods the base which it contains, and which is required to perform the part of a mordant, by preserving to this base all its chemical power, and the physical state most favorable to the reflection of the luminous rays.

CHAPTER XIX.

ALUMINOUS MORDANTS.

The aluminous compounds which are used to deposit on stuffs the oxide of aluminum in the state in which it acts as a mordant, by attracting to it and fixing the coloring matter of a dye-bath, are of two kinds. In some, the alumina is in the state of a base; in others, it performs the part of an acid.

In the basic state, there are as many aluminous salts as acids, but all of them cannot be employed as mordants, those which are insoluble are taken off, by the slightest washing, from the stuff on which they are applied; such are the tri-basic sulphate, the phosphate, the phosphite, the arseniate, the borate of alumina, *et cetera*. Those which are soluble behave in three different manners: some are *basic*, or capable of becoming so by giving up a part of their acid, and therefore require to be only deposited on a fabric to yield to the fibre, either in the cold or with the aid of a temperature more or less elevated, all or part of the alumina which they contain: such are the pure or impure acetate of alumina, cubic alum, oxalate of alumina

ALUMINOUS MORDANTS. 149

the butyrate and the formiate. Others, either neutral or containing an excess of acid, are divided into two groups; 1st, the salts of alumina in which the oxide is not masked, and which, consequently, may always become mordants or yield their oxide to the goods when their acid is saturated with no base, or when, by the aid of another salt, by double decomposition, the formation of a new aluminous salt, insoluble and adherent to the stuff, is determined; to this category belong the sulphate, the seleniate, the chlorate, the bromate, the iodate, the bi-phosphate, the bi-arseniate, the nitrate, the chromate, the chloride, the bromide, the iodide, and octahedral alum; 2d, the salts of alumina of which the base is masked, and which, saturated by an oxide, or mixed with another salt, would never furnish to the fabric an aluminous compound, insoluble, adherent, and capable of attracting the coloring matter. It is in this group that the tartrate, the citrate, and the malate of alumina range themselves. Thus, with the exception of these last three, it may be said that all the compounds of alumina can serve for mordants; with this difference, nevertheless, that some require only to be deposited on the stuff, at a temperature more or less elevated, to fix their base upon it, while others would remain upon it indefinitely without giving up alumina to the fabric, if by the intervention of something the base did not become free and insoluble. This will be better understood by repeating the following experiments of Persoz. After previously scouring with an acid from all foreign matters, the samples of calico, A, B, C, D, E, he impregnated—

Sample A with a solution of acetate of alumina at 6° Twaddell;

Sample B with a solution of nitrate of alumina in the preceding liquor, and marking 12° Twaddell;

Sample C with a solution of nitrate of alumina at 6° Twaddell;

Sample D with a solution of *alum* in an acetate of alumina at 3°, and marking 9° Twaddell;

Sample E with a solution of alum marking 9° Twaddell;

and these samples, dried at the same temperature, in the same conditions, then rinsed several times in distilled water, lastly dyed in a madder bath, were found as follows:—

Sample A, charged with coloring matter of an intensity proportional to the quantity of oxide yielded to the fabric by the acetate.

Sample B—though impregnated with a preparation containing much more alumina—was dyed a much weaker shade, showing the influence of the nitrate which always renders the decomposition of the acetate a little more difficult.

Sample C, always colorless when the nitrate of alumina employed contained one equivalent of base for three equivalents of acid, and the cloth on which it was applied was entirely freed from the calcareous substances with which it is sometimes charged on coming from the operations of bleaching, which are always finished with washings in water.

Sample D, of a shade less intense, by half, than that of sample A, so that the alum associated with the acetate of alumina was a pure loss in the process.

Sample E, colorless like sample C, and in the same conditions.

When other samples, A', B', C', D', E', were impregnated with the same solutions, but after being dried were passed into menstrua containing either bi-carbonate of potash or soda, or the neutral arseniate of potash and a little chalk, or any other saturating body incapable by its nature of redissolving the aluminous compound which is formed; and when, as in the preceding case, all the samples had been washed and passed into a madder bath, the following is the state in which they presented themselves:—

Sample A' had a shade of a much higher tone than sample A.

Sample B' was of a shade double the intensity of that of sample B.

Sample C' of the same shade and tone as sample A', while C was colorless, or very slightly tinted.

Sample D' of a deeper dye than D, intermediate between those of A' and B'.

Sample E', instead of being colorless as sample

E, had a tint the intensity of which was proportional to the alumina of the alum which was fixed.

Chloride of alumina gives the same results as the nitrate.

Oxalate of alumina presents an important peculiarity, which must be taken into consideration; at the moment of its formation it has not the property to yield its basis to the goods, but by prolonged contact, or instantaneously by action of the steam, this salt undergoes a transformation, and giving a part of its basis to the goods, becomes a mordant.

ALUM is of all ingredients the most generally employed, and that which has been longest in use. The octahedral alum has always the property of yielding to the stuff all or part of the alumina it contains, when it has been previously saturated with acetate of lead, lime, baryta, &c., which, by double decomposition, gives sulphates more or less soluble and a proportionate quantity of acetate of alumina.

OLD MORDANTS.

Red mordant, from 1760 to 1800. In 22 gallons of water, they dissolved,

      55.5 lbs. alum, to which they add
      5.5 " arsenious acid,
      5.5 " litharge,
      14.0 " acetate of lead,
      1.54 " sulphuret of antimony,
      1.54 " chloride of mercury,
      3.3 " carbonate of soda.

ALUMINOUS MORDANTS.

Other from 1800 to 1824. In 22 gallons of water, they dissolved,

49.5 lbs. alum, and to this add
5.0 " acetate of copper, previously dissolved in one quart of acetic acid,
27.5 " chlorhydrate of ammonia,
24.2 " carbonate of potash,
24.2 " " lime,
19.1 " acetate of lead.

NEW MORDANTS.

Mr. D. Kœchlin, in his memoir on red mordants, gives the composition of the three following:—

Mordant No. 1.

In 22 gallons of water dissolve
88.0 lbs. alum,
8.8 " carbonate of soda,
88.0 " acetate of lead.

Mordant No. 2.

In 22 gallons of water dissolve
60.0 lbs. alum,
6.0 " carbonate of soda,
44.5 " acetate of lead.

Mordant No. 3.

In 22 gallons of water dissolve
44.5 lbs. alum,
5.0 " carbonate of soda,
29.7 " acetate of lead.

The following is the process to prepare these mordants:—

In a tub containing the powdered alum, pour the quantity of warm water necessary to dissolve it, then add the carbonate of soda, and at last the acetate of lead. A precipitate of sulphate of lead is formed. Shake the whole for one hour without interruption, and afterwards from time to time only. When the mordant has cooled and the sulphate of lead has deposited, decant the clear liquor and keep it in stoneware vessels.

It would seem, at first view, that in all establishments, it must exist a mother-mordant with which all the others might be prepared by diluting it more or less with water, and making additions to it of substances suitable for the different shades; however, it is not the custom of dyers and calico printers who prefer to prepare several kinds of mordants, being guided by the following considerations:—

1. There are a few shades for which a very strong mordant is required, or one demanding a greater quantity of acetate of lead than a mordant of mean density.

2. This last, into the preparation of which less acetate of lead enters, keeps longer than a strong mordant, which soon, by decomposition in the cold, depositing more subacetate of alumina than the mordant of mean density, would not always give a constant result when diluted with water.

3. A strong mordant, in which the acid acetate predominates, would not suit in several styles of printing, especially in that which consists of two or three reds where mordants of different density are printed one on another, because then the mordants getting confounded together would produce less distinct tints.

4. The mode of giving consistence to a mordant, or of thickening, varies according to the kind of printing for which it is intended, and an acid mordant cannot be inspissated so easily as another, with any of the substances which are employed for that purpose.

5. A strong and acid mordant is less easily discharged by the operation of dunging.

In many calico-printing works in the neighborhood of Paris and Rouen, they use for the preparation of the red mordants, sulphate of alumina, which is now manufactured in pretty large quantities. As it occurs in commerce, it contains:—

	Centesimally.
Sulphuric acid	33.178
Oxide of aluminum	17.820
Water	49.002
	100.000

It requires, therefore, seventy-five parts of acetate of lead to effect its partial saturation, and one hundred and eighteen parts of this same salt to render the double decomposition complete, and in order that all the sulphuric acid may be precipitated in the state of insoluble sulphate of lead. Nevertheless, these proportions of acetate may vary considerably, for, as has been already remarked, the composition of the sulphate of alumina is not always the same. It is certain that the commercial article contains different quantities of acid and of base, and the manufacturer cannot exercise too much circumspection in the use of this salt, especially for certain kinds of printing.

M. D. Kœchlin prepares the red mordant with the sulphate of alumina by operating in the following manner:—

To one hundred and ten parts of a solution of sulphate of alumina, marking 52° Twaddell when it is hot, and 56° when cold, he adds one hundred parts of acetate of lead dissolved in thirty parts of water; a double decomposition takes place between these two salts, and a solution of acetate of lead is obtained, marking 24° to 26°—the most concentrated which can be obtained.

There are print-works in which the acetate is replaced by an equal weight of acetate of lead; but when one does not wish to use either the one or the other, equivalent quantities of acetate of lime, baryta, or soda may be substituted, since

2375 pounds crystallized acetate of lead are replaced either by
1600 pounds anhydrous acetate of baryta, or by
1708 " crystallized acetate of soda, or by
1233 " anhydrous acetate of potash

ALUMINOUS MORDANTS.

If commerce supplied the market with the acetates of baryta or lime in a state of purity, the manufacturer would find a great advantage in using them, because he would leave the sulphate of lime or of baryta, the product of the double decomposition, mixed with the mordant, and these salts would contribute as a mastic to the thickening of the color.

Instead of making the mordants by the way of double decomposition, which always necessitates the employment of an acetate, the mordant of which M. D. Kœchlin indicated the preparation has long been manufactured on the large scale, and the following is the process employed: 1. Neutralize a solution of alum, saturated in the cold, with carbonate of potash, which is added by degrees with agitation, till the flakes which are formed begin to be no longer redissolved. 2. Bring this neutralized solution to the boiling point, so as to cause the formation of basic sulphate of alumina, which is collected and afterwards treated with acetic acid, wherein it dissolves perfectly, especially in the heat, furnishing one of the strongest and most reliable mordants that can be prepared and employed. But it would be too troublesome to make this preparation on a small scale and in the works themselves, since it would be necessary to throw away the water from which the basic sulphate of alumina had been separated, and along with this water the sulphate of potash, so that all the potash of the alum, the whole of that which served for its precipitation, and lastly, a certain quantity of the alum itself would be lost. If, on the contrary, the fabrication of this product were conducted on the large scale in an alum factory, where the water more or less saturated with sulphate of potash might enter again continually into a new operation, there would be no loss of alkali; the basic sulphate of alumina produced would be constant in its composition, dissolving well in the acetic acid; and in this case one would economize the whole of the potash of the alum, which might be turned to good account, and all the oxide of lead, when the acetate of this base was employed.

Applications.—The mordants of alumina are employed alone or with some other mordants, for the fixation of all coloring matters, which require an intermediate agent to constitute a color, and to become afterward adherent to the goods.

CHAPTER XX.

FERRUGINOUS MORDANTS.

The ferruginous preparations, like aluminous ones, only perform the part of mordants so far as they are soluble, and cause a deposit of oxide of iron on the stuff. Iron presents several degrees of oxidation, and it is necessary to find, not only the saline combination which best gives up its base to the stuff, but further, that which possesses, in addition to this property, the degree of oxidation necessary to attract the coloring matters without injuring the goods. The fact must not be lost sight of, that, in depositing a ferruginous preparation on the goods, the iron may be combined either in the state of protoxide, which passes by little and little to the state of sesquioxide and even of ferroso-ferric oxide—Fe_3O_4; or in the state of sesqui-oxide, which may be hydrated, namely, in that in which it preserves its chemical condition, or anhydrous, exhibiting that modification in which it is, so to speak, unfit to perform any part; are lastly in the state of a subsalt or insoluble neutral salt.

In a paper entitled, *Employment of pyroligneous acid in some operations of the arts,* and published in the *Annales des arts et manufactures,* M. Bosc examines in what state of oxidation iron should exist on the goods to serve as a base for black. According to this author, one should obtain on cotton a deep black tint, firm and brilliant, only in so far as use is made of a salt of iron with a base of black or protoxide, and the most favorable combination would result from the solution of the iron in acetic acid, because this acid, by the carbon which it contains, would prevent oxidation, and maintain the oxide at its inferior degree.

Arriving at the same conclusions, in a very extended memoir which treats of the fixation of the mordants of iron on cotton goods, M. H. Schlumberger establishes, first, that the acetate of iron obtained by several processes gives results very similar, and bases this proposition on the following experiments:—

He thickened with gum-water on the one hand, and with starch on the other, the following solutions of equal strength—10° Twaddell—*videlicet,*

The *first,* of acetate of iron obtained by the double decomposition of sulphate of iron and acetate of lead.

The *second,* of acetate of iron produced from a solution of iron in acetic acid.

The *third,* of acetate of iron produced by a solution of the metal in ordinary vinegar.

The *fourth*, of acetate of iron prepared by means of partially purified pyroligneous acid.

The *fifth*, of acetate of iron from which the tar had been separated by five minutes' boiling.

The *sixth*, of crude acetate of iron containing a great excess of tar.

The *seventh*, and last, of crude acetate of iron mixed with the purified salt.

These compositions were printed in the same conditions on pieces of calico; each resulting sample was then divided in two, and exposed to the atmosphere, one-half for two days only, the other for ten, before being submitted to the operation of dunging, and passed into a madder-bath where all gave a very fine violet, intense and very rich.

When an acetate is employed as a mordant, theory and practice direct that the proto-acetate of iron be applied, in preference to the goods, and this, by decomposing on the stuff, passes by slow degrees to the state of a basic salt, which oxidizes in the air; and, as it was desirable to inquire into the circumstances in which this oxidation might be effected without danger to the fabric, M. H. Schlumberger turned his attention to the question, and relates the results of experiments which he made on the four ferruginous preparations which follow, some at 24° Twaddell, and others at only 7°.

1. Acetate of iron obtained directly from the solutions of iron in acetic acid.

2. Crude acetate of iron.

3. Acetate of iron obtained by the double decomposition of acetate of lead and sulphate of iron.

4. The same solution, but with an excess of acetate of lead added.

After printing these different solutions, gummed and not gummed, on as many samples as were necessary to study the different circumstances of oxidation, he exposed some, in a place with a mean temperature, to a moist air and diffused light; others in a warm situation, dry and darkened; others in fine to the rays of the sun and to all the atmospheric variations; and left in these different conditions the half of each of these samples for six days, and the other half for twenty-one days; then he passed them all into dung, to be subsequently cleaned and dyed, after which he found—

1. That the weakening of the stuff generally took place only in the samples on which the concentrated ferruginous solutions had been printed, and that in one case only, this weakening was remarked on the stuffs impregnated with a solution marking $6°$;

2. That the goods were weakened by any of the four mordants mentioned above; less, however, with the last, containing an excess of the acetate of lead;

3. That the pure mordants weakened the stuff

much more than those which were thickened with gum, starch, or fecula;

4. That exposure to the solar rays promotes in a given time the injurious effect on the goods, to such a degree that weak mordants, which do not attack the calico in darkness or in a diffuse light, deteriorate it very powerfully in the sun;

5. That in all the cases the weakening of the fabric does not decidedly show itself till the third or sixth day, but that at this period it is nearly the same as after the twenty-first day of the contact of the mordant with the stuff;

6. Lastly, that as the samples are passed into the dung at a boiling heat, or only at the temperature of 122°, and according as, on taking them out of this bath, they are or are not dipped into a dilute solution of chlor-oxide of calcium, the deterioration of the fabric is more or less decided, that is to say, it is scarcely perceptible if the samples have been cleared in a dung-bath heated to 122°, and if they have not been passed into bleaching powder liquor; and, on the contrary, it is always strongly marked when the same samples have been passed into the dung at a boiling temperature, or immersed immediately in the chlor-oxide.

After having thus shown, on the one hand, that this weakening of the fabric is due to the oxidation which takes place by reason of the quantity of protoxide which is deposited upon it, and on the other, that it is reduced to nothing when the mordants are weak, and is very marked when they are concentrated, M. H. Schlumberger explains this by the consecutive effects of the combination of the protoxide with the fabric, a circumstance involving disengagement of heat and electricity. M. Persoz accounts for this phenomenon by the fact of the momentary production of ferric acid—FeO_3—which, as he ascertained by direct experiment, destroys the tissues with great energy when it is free in their presence.

It appears, from the researches of Schlumberger, that if, for fast impressions in black or violet, use is made of crude acetate of iron strongly charged with a tar which obstinately maintains the iron in the state of protoxide on the cloth, very bad results are obtained in the dyeing, whilst the same salt mixed with a certain quantity of acetate, prepared by the solution of iron in acetic acid, never gives any but good results.

To these two orders of facts—which demonstrate, the one, the inefficacy of a mordant too energetically maintained in the state of protosalt, the other, on the contrary, the efficacy of the mordant which is capable of passing to a superior degree of oxidation—Schlumberger adds others, which he adduces as affording unequivocal proof that a too advanced oxidation is always hurtful.

Thus, for example, after having steamed samples

on which were printed mordants of violet and puce-color—mixture of iron and alumina—he remarked that these samples, when dyed and heightened, presented shades of a much more reddish tint than if the mordants had not been submitted to the action of the steam, which, nevertheless, appeared to him more hurtful to the puce mordants containing alumina, than to the black mordants with an iron base, and hence he concluded that this result is due to a more advanced oxidation; but Persoz thinks that there is here a misapprehension as to the part performed by the steam, which does not, in his opinion, set up any phenomenon of oxidation, but simply a change of physical state due to the heat, which renders indifferent a certain quantity of the oxides of iron and aluminum that are fixed on the stuff, and produce in this case, mixed with the violet—the sesqui-oxide, a kind of brown, and alumina, a less full shade.

Other samples, impregnated in like manner with mordants, and dipped, some into a solution of bichromate of potash, others into a bath of bleaching powder diluted and heated to 104°, did not give better results; the tints of the samples passed into the bichromate were even more reddish than those of the specimens passed into the steam, which may be accounted for, when it is borne in mind that always when a stuff on which a protosalt is printed, is dipped into a solution of bichromate of potassa, there is a double decomposition, followed by deterioration, and consequently the formation of a compound which may be represented by a certain quantity of sesqui-oxides of chrome and iron; now, these acting as mordants, and the former producing brown shades, it is not surprising that one cannot obtain fine violets.

As for the action of the chlor-oxide of calcium, it is very simple: it modifies the physical state of the sesqui-oxide without changing its composition.

According to Mr. Mercer, the best iron mordant is the crude acetate—pyrolignite—properly made, free from tar, but containing all the ethereal oils and spirit, as also the deoxidizing coloring matter, which prevent the too rapid oxidation of the iron. This mordant, combined with a proper quantity of white arsenic—arsenious acid—so as to form sesqui-arsenite of iron as oxidation progresses and acetic acid evaporates, is the height of perfection for lilacs and fine plate work. The English purple plate styles from this mordant are unequalled.

To sum up, it may be affirmed, without fear of contradiction from experiment, that when solutions of iron obtained by acetic acid are applied on the stuff, with the view of making them perform the part of mordants, it is right that they be in the state of protoxide, in order that, the oxidation taking place on the cloth, there may be formed a basic acetate which will preserve to the sesqui-oxide its chemical properties, and pass to the state of phosphate or arseniate in the operation of dunging. It is necessary that this oxidation be slow and progressive, for, if it is rapid, the risk of the stuff being deteriorated, or of the sesqui-oxide passing into that isomeric state in which it becomes, as it were, indifferent to chemical agents, is incurred.

As for the other ferruginous compounds, all the acid salts are unfit to perform the part of mordants, while it is otherwise with the neutral salts, seeing that the protoxide which they contain, passing to the state of sesqui-oxide by absorbing the oxygen of the air, they no longer contain enough of acid to form a neutral salt, and consequently there is the formation of a basic salt which becomes fixed on the stuff. It is thus that one explains why the neutral protosulphate which remains on the calico yields to it always a certain quantity of its base, whereas, when it is acid, this phenomenon no longer presents itself. As for the sesquisalts, all those which, from any cause whatever, can pass into the state of basic salts, then become true mordants, capable of attracting coloring matters.

When the iron is in contact with the calico in presence of moist air, it produces, by oxidizing, spots of rust, which become fixed on the cloth and attract the coloring matter.

In the same circumstances the protosulphate presents the same results, either, from the circumstances that, passing into the state of sulphate, by an absorption of oxygen, it is immediately transformed into a basic salt by fixing a higher proportion of oxygen, or that it has directly the power of fixing by a double decomposition a certain quantity of coloring matter.

The Alkaline Mordants of Iron.—Few, besides *Haussmann*, have employed as mordants alkaline ferruginous solutions. He dissolves iron, or the protosulphate gently in nitric acid, under which condition there was always the formation of an ammoniacal salt. The following directions explain the reaction.

$$8Fe + 19NO^5 + 4HO = 5(Fe^7O^3, 3NO^5) + NH_3NO^5HO +$$

Iron. Nitric acid. Water. Sesqui-nitrate of iron. Nitrate of ammonia.

$$NO^2 + NO$$

Bioxide of nitrogen. Protoxide nitrogen.

The liquor abstracted was afterwards saturated with carbonate of potash, which was poured in very cautiously. The precipitate which formed at first was soon redissolved by an excess of carbonate of potash, giving rise to a double salt, which was decomposed by the alkaline oxide. *Haussmann* states that he uses this solution with success in many circumstances.

Applications.—Those mordants are used alone or mixed with those of alumina. In the first case they serve with the red coloring matters to produce on the stuffs gray, lilacs, violets and blacks; with the yellow they give grays, olives more or less deep, with a mixture of red and yellow, a multitude of shades, from clear gray to the deepest black. Associated with alumina mordants, the ferruginous give with red coloring matters, pure shades more or less intense; with yellow, yellows more or less olive; with a mixture of red and yellow they give brown colors, dead leaves, rather mauve, &c., which vary indefinitely according to the respective proportions of the mordants of alumina and iron.

CHAPTER XXI.

STANNIFEROUS MORDANTS.

Tin, by uniting with oxygen, gives two oxides, one which reacts as a powerful base, the other as an acid; both are applicable as mordants. From all metallic compounds the stanniferous combinations are those which adhere to the goods with the greatest energy. The choice between a stannous and stannic salt is determined by the nature of the goods, and by that of the colors that it is desired to fix upon them. It will here be sufficient to consider the conditions in which these compounds must exist.

The compounds in which the oxide of tin performs the part of a base are of two kinds; some having a base of protoxide, and others of binoxide. The protoxide is the most generally used; it cannot be separated on the stuff without giving up to it a certain quantity of its base, seeing that, when treated with water, it undergoes a partial decomposition, and is transferred into an acid salt, which remains in solution in that medium, and

STANNIFEROUS MORDANTS.

into a basic insoluble compound, which adheres to the fabric.

Instead of chloride of tin, BANCROFT employed a solution of the protosulphate in hydrochloric acid, which decomposes more easily in presence of the goods. He prepares it in the following manner: On 22 lbs. of granulated tin, introduced in a stoneware vessel, he pours 36 lbs. of commercial hydrochloric acid free from iron, adds little by little to this mixture $16\frac{1}{2}$ lbs. of sulphuric acid; there is development of heat, the tin is attacked first with violence, but it dissolves more slowly in proportion as the liquor comes more concentrated. The mixture is heated in a sand bath till complete dissolution. The whole being left to cool, a saline mass is obtained which contains a slight excess of tin. The liquor is decanted, the remaining metal is weighed to know how much has been dissolved, and the liquor is diluted with as much water that its weight may be eight times that of the tin dissolved; that is to say, 160 lbs. for example, if there have been 20 lbs. of tin dissolved. Among the compounds of binoxide of tin, there is a multitude of preparations which are employed as mordants or constituent parts of the latter which are applied on the goods, and which contained binoxide, either pure or mixed with protoxide. They are generally called *Tin compositions*. The following are some of them:—

1st. 22 lbs. of tin in ribands are dissolved with precaution in a mixture of
    55 " nitric acid,
   120 " commercial hydrochloric acid.

2d. 22 " granulated tin are dissolved in a mixture formed of
    44 " hydrochloric acid,
    44 " nitric acid in which has been previously dissolved
    11 " hydrochlorate of ammonia.

3d. 22 " tin in ribands are gradually dissolved in
   176 " nitric acid at 40° in which has been previously dissolved
    22 " chlorhydrate of ammonia.

4th. 22 " tin are dissolved in
    22 " nitric acid at 62°,
    44 " hydrochloric acid,
    44 " water.

5th. 22 " protochloride of tin are dissolved in a mixture of
    35 " hydrochloric acid and
    $17\frac{1}{2}$ " nitric acid;

or of
    $17\frac{1}{2}$ " hydrochloric acid or
    15 " nitric acid;

or lastly of
    11 " hydrochloric acid,
    15 " nitric acid.

6th. 22. lbs protochloride of tin are gradually dissolved in
27½ " nitric acid.

Further, in a mixture formed,
7th. 22 lbs. nitric acid, and
22 " hydrochloric acid,

as much tin is dissolved as these acids can reduce, and then heat is applied to dissolve in this liquor previously decanted
2.2 lbs. protochloride of tin.

8th. 22 " tin are dissolved with caution in
42 " nitric acid at 64°,
33 " hydrochloric acid at 36°; the solution being effected, add
5½ " acetate of lead.

Lastly, protochloride of tin is dissolved by small portions at a time to the point of saturation in nitric acid at 66° or 68°. The resulting solution has the consistence of a jelly.

With reference to the Compounds in which Oxide of Tin performs the part of an Acid.

These mordants are of frequent use; they are prepared by dissolving protoxide of tin, or, for greater economy, protochloride, in hydrate of potash or soda. These bases form with chlorine alkaline chlorides, and the stannous acid set free, combines with the excess of base to form a soluble stannite.

This compound has very little stability; carbonic acid of the air tends to decompose it, and another cause is the unstability of its molecules, the atom of protoxide is divided in two and is transformed in binoxide, and metallic tin like shows the following reaction:—

$$2SnO = SnO^2 + Sn$$

Application of the tin mordants.—The tin mordants are rarely employed to obtain *dyed* colors on those called *maddered;* they are used to combat the effects of iron, or after the dyeing is effected, to transform by substitution a lake with a base of alumina into another lake with a stanniferous lake. These mordants figure in all the colors of *application*, and specially in *steam colors*.

Other mordants are used to fix colors on fabrics as compounds with a base of sesquioxide of chrome. But although the latter oxide is isomorphous with alumina, and sesquioxide of iron is susceptible of adhering to the goods, and attracting coloring matters, it gives rise, by its greenish gray shade, to lakes which are not clear in the colors.

These compounds, as well as those of some other metallic oxides, not being in general use, do not require to be minutely discussed, and with reference to the fatty organic mordant which plays so important a part in the *Turkey red*, we refer to general works on the art of dyeing and calico printing.

CHAPTER XXII.

ARTIFICIAL ALIZARIN.

WE have seen that the bi-nitro-naphthaline is a fecund spring of colored products; the action of reducing agents, such as the sulphurets, the stannous salts dissolved in caustic potash, the cyanide of potassium, &c., give with this substance derivated products which are red, violet, blue and very rich. When the reducing agents are of an acid nature, such as a mixture of zinc and diluted sulphuric acid, iron filings and acetic acid, the bi-nitro-naphthaline is not altered.

If you make to act concentrated sulphuric acid on the crystallized bi-nitro-naphthaline, it is no chemical reaction. In heating the mixture at 482° the bi-nitro-naphthaline is dissolved, and sulphuric acid begins to act only after a long ebullition. When this solution is diluted with water, the bi-nitro-naphthaline is precipitated unaltered; the same if you treat madder at 212°, by concentrated sulphuric acid all the products are destroyed but one—the coloring principle—or *alizarine*.

The formula of alizarine is represented by—
$$C^{20}H^6O^6$$
that of the bi-nitro-naphthaline by—
$$C^{20}H^6(NO^4)^2$$

A reducing agent capable to take two molecules of oxygen, and change the nitrogen in ammonia, could probably change the bi-nitro-naphthaline in alizarine, and the experiment has confirmed that theory. The following process permits to prepare artificial alizarine:—

Make a mixture of bi-nitro-naphthaline and concentrated sulphuric acid, that you introduce in a large dish heated in a sand bath. By the action of heat the bi-nitro-naphthaline is dissolved in sulphuric acid. When the temperature is at about 392°, throw in it some small pieces of zinc; few minutes after it disengages sulphurous acid; half an hour after the operation is achieved. If you let fall a drop of the acid mixture in cold water, a magnificent red violet color is formed, due to the formation of artificial alizarine; sometimes the reaction is very energetic; if the quantity of zinc is too considerable, the sulphuric acid boils rapidly, and a large quantity of white vapors are disengaged. The zinc must be added by small portions at a time.

When the reaction is achieved, dilute the liquid with eight or ten times its volume of water, and boil, few minutes after you filter. The artificial alizarine deposit on form of a red jell. The other water is strongly colored in red, and contains a considerable quantity of alizarine in solution. This water can be used directly to dye.

In the preceding reaction the zinc can be substituted by many other substances, such as tin, iron, mercury, sulphur, carbon, &c. The two following equations show the reaction:—

$$\underbrace{C^{20}H^6(NO^4)^2}_{\text{Bi-nitro-naphthaline.}} + 12M + 18SO^3,HO = \underbrace{C^{20}H^6O^6}_{\text{Alizarine.}} + \underbrace{2(SO^3NH^3HO)}_{\text{Sulphate of ammonia.}}$$

$$+ \underbrace{12SO^1MO}_{\text{Metallic Sulphate.}} + 10HO + 4SO^2$$

$$\underbrace{C^{20}H^6(NO^4)^2}_{\text{Bi-nitro-naphthaline.}} + 10C + 14(SO^3HO) = \underbrace{C^{20}H^6O^6}_{\text{Alizarine.}} +$$

$$2(SO^3,NH^3HO) + 10CO^2 + 12\,So^2 + 6HO$$

In the first equation it is the metal which acts on sulphuric acid, in the second it is the carbon itself.

This artificial alizarine has all the characters of the ordinary alizarine. The following table shows how the two coloring matters comport:—

COLORING MATTER OF THE MADDER	ARTIFICIAL RED MATTER
is precipitated in jell from its solutions,	is precipitated in jell from its solutions,
is sublimated between 419° and 464°.	is sublimated between 419° and 464°.
Little soluble in water, soluble in alcohol, ether, and a solution of alum,	Little soluble in water, soluble in alcohol, ether, and a solution of alum,
unalterable by sulphuric acid heated at 392°, hydrochloric acid; alterable by nitric acid,	unalterable by sulphuric acid heated at 392°, hydrochloric acid; alterable by nitric acid,
soluble in caustic or carbonated alkalies with a purple color.	soluble in caustic and carbonated alkalies with a blue violet color.
The ammoniacal solution gives purple precipitates with the salts of baryta and lime.	The ammoniacal solutions give purple precipitates with the salts of baryta and lime.

The elementary analysis gives—

Carbon . . 63.26 . . 63.51
Hydrogen . . 2.10 . . 2.30

New studies deserve to be done on this interesting body, which is called to render important services in the arts of dyeing and calico printing.

This new substance gives colors as good and solid as the carmine of madder for impression and fixation of colors by steam on mordanted cotton cloths.

CHAPTER XXIII.

METALLIC HYPOSULPHITES AS MORDANTS—DYER'S SOAP—PREPARATION OF INDIGO FOR DYEING AND PRINTING—RELATIVE VALUE OF INDIGO—CHINESE GREEN—MUREXIDE.

Mr. E. Kopp, a short time ago, has introduced the use of metallic hyposulphites as mordants, and he has shown that their use is preferable to the acetate of the same base. The hyposulphite of lime is the one used to obtain the others, its fabrication is known by every chemist.

Hyposulphite of Alumina.

To prepare a solution of hyposulphite of alumina he decomposes 64.60 grains of sulphate of alumina $(3(SO^3)Al^2O^3+18HO)$, dissolves in water by 75.66 grains of crystallized hyposulphite of lime; he filters and expresses the residue of sulphate of lime. The solution is clear, limpid, and kept very well to the air; a solution of hyposulphite of alumina marking 1.20 contains as much alumina as a solution of acetate of alumina at 1.10 of specific gravity. This solution can be thickened by gum, roasted starch, &c.

If you use alum, you find that 13½ lbs. of alum are decomposed completely by 9 lbs. 2 ounces of hyposulphite of soda ($S^2O^2NaO+5HO$), or by 9 lbs. 3 ounces of crystallized hyposulphite of lime ($S^2O^2CaO+6HO$). It follows that 4½ lbs. of this last salt can take the place of 6 lbs. 12 ounces of acetate of lead.

Hyposulphite of Protoxide of Iron.

This salt can be obtained by the action of sulphurous acid on protosulphuret of iron mixed with water, or by the decomposition of the protosulphate of iron by the hyposulphite of lime; it must be kept out of the contact of the air. In dyeing it behaves like the other iron mordants.

Hyposulphite of Chrome.

This salt is prepared like the corresponding salt of alumina. It must be prepared a short time before its use.

Hyposulphite of Tin.

All stannous salts being acids when they are mixed with an alkaline hyposulphite, they disengage hyposulphurous acid. With the salts of protoxide of tin, they form a stannous sulphuret or oxy-sulphuret which are precipitated with stannoso-stannic; this formation takes a certain time according to the concentration of the liquors.

In using a salt of peroxide of tin, it is no precipitation of tin in the liquors. The above observation shows that in using hyposulphites, you must avoid mixing with a stannous salt, but always use a stannic salt or a mixture of them both.

This salt gives a very good mordant.

SOAP FOR DYERS.

This soap is composed of—

Ordinary oil or fatty body	180 lbs.
Palm oil	11¼ "
Spirit of Turpentine	33¾ "
In all	225 "

Dyers add to it from 5 to 6 quarts of lye of potash at 5°, and 18 to 20 quarts of lye at 22°. The coction of the soap lasts twelve hours.

PREPARATION OF INDIGO FOR DYEING AND CALICO PRINTING.

TAKE 2 qts. of a paste containing about 2 lbs. of indigo in fine powder, mix with it 2 qts. of glucose prepared with rice starch. Take afterwards 2½ lbs. of slacked lime diluted with water, that you mix with the glucose and indigo, add then 2 lbs. of solid caustic soda and shake carefully. This compound thus prepared is ready for impression which is executed by the ordinary process. To dye with this indigo mix together the materials, viz: indigo, glucose, lime, soda, let it work a certain length of time at the ordinary temperature, and introduce it in the vat ready for the dyeing.

Relative value of indigo.

Country.	Relative value in coloring matter.	Ashes in 100 parts.	Water in 100 parts.
Indigo of East India	68.	4.5	5.0
" "	66.	5.8	6.0
" "	64.	8.1	8.0
" "	54.	11.0	7.0
" "	51.5	7.2	7.5
" "	54.	3.6	7.0
" "	45.	14.0	8.4
Spanish Indigo	55.	12.3	6.0
" "	50.	13.0	7.0
" "	44.5	19.0	5.5
" "	28.	33.4	4.5
Bengal	64.	5.9	4.0
"	47.	24.6	5.0

Benares	45.	20.7	8.4
Guatemala	50.	16.0	6.5
Madras	41.	10.6	6.7
Oude	46.	6.3	8.5
Caraccas	52.5	16.2	6.4
Madras	35.	33.3	6.0
Java	63.5	5.4	4.8
Bengal	59.5	7.5	5.0
"	56.	11.0	5.3
"	45.5	14.0	7.2
"	24.	44.0	4.4
Manille	35.5	28.0	5.0

China Green.

Mr. Charvin has extracted from the *Rhamnus catharticus* a green coloring matter similar to the Chinese green (green indigo) but less costly. This product is in irregular plates with a variable aspect, according to the thickness of the plate. Like the Chinese *Lo-Kao* this product seems to be a lake, that is, a combination of an organic substance with an earthy matter. Gradually heated, it lost first water without any sublimate product; in burning, it left a considerable quantity of ashes. The following is the result of a comparative experiment done at the same time with that product and the *lo-Kao*, with the analysis of Mr. Persoz :—

	Green. Charvin.	Chinese.	Chinese by Persoz.
Water	13.5	9.5	9.3
Ashes	33.	28.5	28.8
Coloring matter	53.5	62.	61.9
	100.0	100.0	100.0

Mr. Persoz defines the *lo-Kao* "a lake formed by cyanine, having for base phosphated magnesia, alumina, and oxide of iron." In Mr. Charvin's process, lime is only found, mixed with a little alumina and silica without phosphoric acid, but the coloring matter is the same in the two products. The chemical reactions of Mr. Charvin's green are similar to the Chinese *lo-Kao*.

Preparation.—In a kettle containing boiling water he puts 2 pounds of *Rhamnus catharticus bark;* a few minutes after a pink skim is produced. He then puts the whole into an earthen jar, well covered, and then allows it to rest till next day. The liquid is yellowish; it is decanted and lime water added to it, which produces a change of color; it turns reddish-brown, the liquid is put in jars—very little in each one—and the whole is exposed to air and light. The reddish-yellow color is modified and takes a green shade; little by little the green color becomes more general, and is then deposited in plates. All the liquids are mixed together and carbonate of potash is added; a green precipitate is produced; he leaves it to deposit, decants the liquid and collects the precipitate and dries it. The experiments of Mr. Charvin prove,

1st. That his green coloring matter is of the same nature as the Chinese *lo-Kao*, and will dye silk in as beautiful a green as the *lo Kao*.

2d. This matter is extracted from an indigenous plant, the *Rhamnus catharticus*.

3d. That the process will permit to manufacture it for dyers at the price of $8.90 per pound.

MUREXIDE.

Murexide can be manufactured with guano or uric acid, the processes are different.

Fabrication with Guano.

The choice of a good guano is important, the one containing the most urate of ammonia is the best. In the best Peruvian guano we found at least 5 and the most 15 per cent. of uric acid.

1. Treat the guano by hydrochloric acid to decompose the carbonate and oxalate of ammonia, the carbonate and phosphate of lime, the phosphates of ammonia and magnesia. This operation is done in a lead kettle. You heat the acid which marks 12° B. and you throw in it gradually the guano by small portions.

2. Boil the mixture one hour, draw the liquid in wooden vessels, wash the deposit by decantation.

3. The guano after this treatment is thrown on large filters; the product thus obtained contains from 42 to 45 per cent. of dry substance.

4. It is in this product that exists the uric acid

mixed with sand, gypsum, organic deposits, and extractive matters.

5. In a porcelain dish put 6 pounds of this guano thus prepared with 1½ pound of hydrochloric acid at 24° B., carry the whole at 122°. Take the dish from the fire and pour in it little by little in shaking all time 7 ounces of nitric acid at 40° B.; be careful that the temperature does not rise above 143° and fall below 111°.

6. The mixture is then diluted with an equal volume of water and filtered, wash the deposit with water, reunite all the solutions and precipitate by a saturated solution of chloride of tin.

7. When the precipitate is well formed, decant the brown liquid and wash it with water containing hydrochloric acid.

8. Throw the precipitate on a filter, dry it and expose it to vapors of ammonia which transform it into murexide.

Preparation of Uric Acid contained in Guano.

The guano is treated by hydrochloric acid as we have seen above.

In a copper kettle of about 125 gallons put 96 gallons of water, 10 lbs. of caustic soda, and the mass obtained by the treatment of 252 lbs. of guano by hydrochloric acid and well washed with water.

Heat the mixture till boiling, shake all time and kept at this temperature for one hour.

Add to it a milk formed with 2½ lbs. or 3½ lbs. of caustic lime; shake well, boil ¼ of an hour, take the kettle from the fire and let it settle for 3 or 4 hours.

Decant, and in the clear liquid put some hydrochloric acid to precipitate the uric acid. Wash this precipitate by decantation, collect it on a filter and dry it.

When you have taken all the clear liquid from the kettle, put on the residue a quantity of water equal to the first used, add again from 6¼ to 7½ lbs. of caustic soda; operate as above except that for the clarification you use only from 19 ounces to 1¼ lb. of lime.

After this second treatment the guano is generally free of uric acid; however, it is good and safe to repeat the operation a third time with less soda and lime.

The uric acid, such as it is, can be used immediately to prepare *Murexide*.

Fabrication of Murexide after the Extraction of Uric Acid.

For 2 lbs. of uric acid you must use 2 lbs. 10 ounces of nitric acid at 36° B.

The acid is put in a dish which is kept in cold water; then you throw the uric acid by portions in the nitric acid; the dose must not exceed one ounce at a time; you must distribute all the uric acid in the mass with a porcelain spatula, and you must not add uric acid till the mixture has come at 80°.

When all the uric acid has been added, you let cool, and then you heat the whole slowly in a sand bath; when the liquid begins to swell take out from the fire, and when the swelling has fallen back begin again. When you heat for the third time, raise the temperature at 230°, and then put in the bath 9¼ ounces of liquid ammonia at 24°B., which transforms the mixture into *murexide*. Leave the dish about 2 minutes on the sand bath, take it out and leave to cool; you found a kind of paste in the mixture which is known by the name of *murexide in paste.*

To obtain it dry and pure, mix that paste with water, filter and wash well; the last washing must be done with ammonia diluted with water; dry in the oven the product left on the filter—it is the *Dry murexide.*

Application.—Murexide can be applied for calico printing in powder or in paste.

Impression with the Color.—In 9 gallons of boiling water dissolve 25½ lbs. of crystallized nitrate of lead, let cool the liquid till 144°; dissolve first in it 5 lbs. of powdered murexide or 15 lbs. of murexide in paste, then 39 lbs. of powdered gum; when all is cold it can be used. The printing terminated, hang the stuffs in a damp place and you fix the purple by ammoniac gas, the same process you pass woollen stuffs to sulphurous acid.

Passage of the Stuffs in the Bath of Sublimate.—
The warm bath in which the tissues are passed after exposition to the ammoniacal gas is composed of 191 gals. of water and 2 lbs. 11 ounces of corrosive sublimate. The tissues are passed in this bath and then in running water, then they receive the bath of acetate of soda.

Acetate of Soda.—This bath is composed of 360 gals. of water with 1 lb. of acetate of soda and 1 lb. of chlorhydrate of ammonia; the tissues are passed in for 20 minutes, then well washed and dried.

This progress gives a very beautiful purple red, all the gradations of red and rose can be obtained with murexide—the colors obtained are very solid.

INDEX.

A

	PAGE
Aluming	44
Aniline, history of	60
properties of	60, 61
direct preparation of	60, 64
artificial preparation of	68
di-nitro	99
green	88
purple	81
to dye with	113
green, to fabrics, method of application of	118
oxalate	67
Allyle-toluidine	82
Art of dyeing, historical notice of the	25
chemical principles of the	33
Azuline	110
Alkaline mordants of iron	168
Alizarin, artificial	175
Alumina, hyposulphite of	179
Acetate of soda	188

B

Benzole, preparation of	68, 69
properties of	68, 71
properties of the bi-nitro	68, 73
bi-nitro	74

	PAGE
Bleaching silk	37
cotton	39
Bleu de Paris	89
Benzolic acid, sulpho-	72
Boiling cotton	40
silk	37, 38
Bath of sublimate, passage of stuff in	189

C

	PAGE
Carminaphtha	106
Chloroxynaphthalate of ammonia	105
Calico with coal tar colors, printing	116
Chloroxynaphthalic acid	104
Coal tar, on the coloring matters produced by	49
history of the coloring matters produced by	49
distillation	52
to the arts of dyeing and calico printing, application of	112
colors, printing with	116
Chloraniline, tri-	62
Chlorophenic acid, tri-	62
Chloranile	62
Cotton	33
Cotton with colors of coal tar, to dye	114
Cotton with molybdic acid, to dye	129
Crysammic acid	125
preparation	125
Cumidine	65, 98
Chrome, hyposulphite of	180
China green	183

D

	PAGE
Distillation and rectification of coal tar, table of the products obtained by the	55
Dyeing	47

E

	PAGE
Emeraldine	88

F

Fibres, preparation of the textile	33
Fixation of coloring matters in dyeing and printing, theory of the	133
Futschine	92
by action of bichloride of tin on aniline, preparation of	93
by action of nitrate of mercury, preparation of	94

G

Guano, preparation of uric acid in	186

H

Hyposulphite, metallic, as mordants	179
of alumina	179
of protoxide of iron	180
of chrome	180
of tin	180

I

Iodaniline	97
Improvements in the art of dyeing	125
Iron, the alkaline mordants of	168
Indigo, preparation of, for dyeing and calico printing	182
relative value of	182

L

Light on coloring matters from coal tar, action of	120
Lutidine	65

M

	PAGE
Madder	130
extract of	130, 131
Magenta	92
Molybdic acid	127
Mordants	43
aluminous	148
old	152
new	153
ferruginous	159
principles of the action of the most important	144
stanniferous	170
Metallic hyposulphites as mordants	179
Murexide	185
fabrication of, after extraction of	185
Uric acid	185
application of	185

N

	PAGE
Naphthamein	109
Nitro-phenisic acid, tri-	64
Nitro-benzole, preparation	68
properties	68, 73
into aniline, transformation of	68
by sulphide of ammonium, reduction of	76
by nascent hydrogen, reduction of	77
by acetate of iron, reduction of	79
Nitro-azo-phenylamine	99
Ninaphthalamine	106
Nitroso-phenyline	98
Nitro-phenyline diamine	99
Nitroso-naphthaline	107
application of	118
Nitroso-phenyline	74

INDEX.

P

	PAGE
Perchloroxynaphthalic acid	104
Picric acid	64, 99, 127, 129
Picoline	65
Preparation of nitro-benzole	73
of binitro-benzole	73
of futschine	93, 94
Pyrrol	65
Pyrrhidine	65
Protoxide of iron, hyposulphite of	180

Q

Quinoline	65

R

Red, tar	110
Roseine	87
to dye with	113
Rosolic acid	101

S

Scouring wool	35
Silk	37
Silk	37
Silk and wool with coal tar colors, dyeing	112
with futschine, picric acid, chinoline blue and violet, to dye	113
with azuline, to dye	114
with molybdic acid, to dye	128
Singing cotton stuffs	39
Stuffs, preliminary preparations of	39
Stannous salt	170
Stannic salt	170
Soap for dyers	181
Sublimate, passage of stuffs in bath of	189
Soda, acetate of	189

T

	PAGE
Tar red	110
Transformation of nitro-benzine into aniline . .	76
Toluidine	65, 98
Tin	170
oxide of, as a base	170
protosulphate of	171
compositions of	171
oxide of, as an acid	173
mordants, application of	174
hyposulphite of	180

U

Ungumming silk	37
Uric acid in guano, preparation of	186

V

Violine	86
to dye with	113

W

Wool	34
Wool with aniline purple, violine, roscine, futschine, to dye	114
with chrysammic acid, dyeing	126

X

Xylidine	98

Practical and Scientific Books,

PUBLISHED BY

HENRY CAREY BAIRD,

INDUSTRIAL PUBLISHER,

No. 406 Walnut Street,
PHILADELPHIA.

☞ Any of the following Books will be sent by mail, free of postage, at the publication price. Catalogues furnished on application.

American Miller and Millwright's Assistant:
A new and thoroughly revised Edition, with additional Engravings. By WILLIAM CARTER HUGHES. In one volume, 12 mo., ...$1.00

Armengaud, Amoroux, and Johnson.
THE PRACTICAL DRAUGHTSMAN'S BOOK OF INDUSTRIAL DESIGN, and Machinist's and Engineer's Drawing Companion; forming a complete course of Mechanical Engineering and Architectural Drawing. From the French of M. Armengaud the elder, Prof. of Design in the Conservatoire of Arts and Industry, Paris, and MM. Armengaud the younger, and Amouroux, Civil Engineers. Rewritten and arranged, with additional matter and plates, selections from and examples of the most useful and generally employed mechanism of the day. By William Johnson, Assoc. Inst. C. E., Editor of "The Practical Mechanic's Journal." Illustrated by fifty folio steel plates and fifty wood-cuts. A new edition, 4to.,....$7.50

Among the contents are:—*Linear Drawing, Definitions and Problems*, Plate I. Applications, Designs for inlaid Pavements, Ceilings and Balconies, Plate II. Sweeps, Sections and Mouldings, Plate III. Elementary Gothic Forms and Rosettes, Plate IV. Ovals, Ellipses,

PRACTICAL AND SCIENTIFIC BOOKS.

Parabolas and Volutes, Plate V. Rules and Practical Data. *Study of Projections,* Elementary Principles, Plate VI. Of Prisms and other Solids, Plate VII. Rules and Practical Data. *On Coloring Sections, with Applications*—Conventional Colors, Composition or Mixture of Colors, Plate X. *Continuation of the Study of Projections*—Use of sections—details of machinery, Plate XI. Simple applications—spindles, shafts, couplings, wooden patterns, Plate XII. Method of constructing a wooden model or pattern of a coupling, Elementary applications—rails and chairs for railways, Plate XIII. *Rules and Practical Data*—Strength of material, Resistance to compression or crushing force, Tensional Resistance, Resistance to flexure, Resistance to torsion, Friction of surfaces in contact.

THE INTERSECTION AND DEVELOPMENT OF SURFACES, WITH APPLICATIONS.—*The Intersection of Cylinders and Cones,* Plate XIV. *The Delineation and Development of Helices, Screws and Serpentines,* Plate XV. Application of the helix—the construction of a staircase, Plate XVI. The Intersection of surfaces—applications to stop-cocks, Plate XVII. *Rules and Practical Data*—Steam, Unity of heat, Heating surface, Calculation of the dimensions of boilers, Dimensions of firegrates, Chimneys, Safety-valves.

THE STUDY AND CONSTRUCTION OF TOOTHED GEAR.—Involute, cycloid, and epicycloid, Plates XVIII. and XIX. Involute, Fig. 1, Plate XVIII. Cycloid, Fig. 2, Plate XVIII. External epicycloid, described by a circle rolling about a fixed circle inside it, Fig. 3, Plate XIX. Internal epicycloid, Fig. 2, Plate XIX. Delineation of a rack and pinion in gear, Fig. 4, Plate XVIII. Gearing of a worm with a worm-wheel, Figs. 5 and 6, Plate XVIII. *Cylindrical or Spur Gearing,* Plate XIX. Practical delineation of a couple of Spur-wheels, Plate XX. *The Delineation and Construction of Wooden Patterns for Toothed Wheels,* Plate XXI. *Rules and Practical Data*—Toothed gearing, Angular and circumferential velocity of wheels, Dimensions of gearing, Thickness of the teeth, Pitch of the teeth, Dimensions of the web, Number and dimensions of the arms, wooden patterns.

CONTINUATION OF THE STUDY OF TOOTHED GEAR.—Design for a pair of bevel-wheels in gear, Plate XXII. Construction of wooden patterns for a pair of bevel-wheels, Plate XXIII. *Involute and Helical Teeth,* Plate XXIV. *Contrivances for obtaining Differential Movements*—The delineation of eccentrics and cams, Plate XXV. *Rules and Practical Data*—Mechanical work of effect, The simple machines, Centre of gravity, On estimating the power of prime movers, Calculation for the brake, The fall of bodies, Momentum, Central forces.

ELEMENTARY PRINCIPLES OF SHADOWS.—*Shadows of Prisms, Pyramids and Cylinders,* Plate XXVI. *Principles of Shading,* Plate XXVII. *Continuation of the Study of Shadows,* Plate XXVIII. *Tuscan Order,* Plate XXIX. *Rules and Practical Data*—Pumps, Hydrostatic principles, Forcing pumps, Lifting and forcing pumps, The Hydrostatic press, Hydrostatical calculations and data—discharge of water through different orifices, Gaging of a water-course of uniform section and fall, Velocity of the bottom of water-courses, Calculation of the discharge of water through rectangular orifices of narrow edges, Calculation of the discharge of water through overshot outlets, To determine the width of an overshot outlet, To determine the depth of the outlet, Outlet with a spout or duct.

APPLICATION OF SHADOWS TO TOOTHED GEAR, Plate XXX. *Application of Shadows to Screws,* Plate XXXI. *Application of Shadows to a Boiler and its Furnace,* Plate XXXII. *Shading in Black—Shading in Colors,* Plate XXXIII.

THE CUTTING AND SHAPING OF MASONRY, Plate XXXIV. *Rules and Practical Data*—Hydraulic motors, Undershot water wheels, with plane floats and a circular channel, Width, Diameter, Velocity, Number and capacity of the buckets, Useful effect of the water wheel, Overshot water wheels, Water wheels with radial floats, Water wheel with curved buckets, Turbines, *Remarks on Machine Tools.*

PUBLISHED BY HENRY CAREY BAIRD.

THE STUDY OF MACHINERY AND SKETCHING.—Various applications and combinations: *The Sketching of Machinery*, Plates XXXV. and XXXVI. *Drilling Machine; Motive Machines;* Water wheels, Construction and setting up of water wheels, Delineation of water wheels, Design for a water wheel, Sketch of a water wheel; *Overshot Water Wheels. Water Pumps*, Plate XXXVII. *Steam Motors;* High-pressure expansive steam engine, Plates XXXVIII., XXXIX. and XL. *Details of Construction; Movements of the Distribution and Expansion Valves; Rules and Practical Data*—Steam engines: Low-pressure condensing engines without expansion valve, Diameter of piston, Velocities, Steam pipes and passages, Air-pump and condenser, Cold-water and feed-pumps, High-pressure expansive engines, Medium pressure condensing and expansive steam engine, Conical pendulum or centrifugal governor.

OBLIQUE PROJECTIONS.—Application of rules to the delineation of an oscillating cylinder, Plate XLI.

PARALLEL PERSPECTIVE.—Principles and applications, Plate XLII.

TRUE PERSPECTIVE.—Elementary principles, Plate XLIII. Applications—flour mill driven by belts, Plates XLIV. and XLV. Description of the mill, Representation of the mill in perspective, Notes of recent improvements in flour mills, Schiele's mill, Mullin's "ring millstone," Barnett's millstone, Hastie's arrangement for driving mills, Currie's improvements in millstones; *Rules and Practical Data*—Work performed by various machines, Flour mills, Saw mills, Veneer-sawing machines, Circular saws.

EXAMPLES OF FINISHED DRAWINGS OF MACHINERY.—Plate A, Balance water-meter; Plate B, Engineer's shaping machine; Plate C D E, Express locomotive engine; Plate F., Wood planing machine; Plate G, Washing machine for piece goods; Plate H, power loom; Plate I, Duplex steam boiler; Plate J, Direct-acting marine engines.

DRAWING INSTRUMENTS.

Barnard (Henry). National Education in Europe:

Being an Account of the Organization, Administration, Instruction, and Statistics of Public Schools of different grades in the principal States. 890 pages, 8vo., cloth,..$3.00

Barnard (Henry). School Architecture.

New Edition, 300 cuts, cloth,..........................$2.00

Beans. A Treatise on Railroad Curves and the Location of Railroads.

By E. W. Beans, C. E. 12mo. (In press.)

Bishop. A History of American Manufactures,

From 1608 to 1860; exhibiting the Origin and Growth of the Principal Mechanic Arts and Manufactures, from the Earliest Colonial Period to the Present Time; with a

Notice of the Important Inventions, Tariffs, and the Results of each Decennial Census. By J. Leander Bishop, M. D, : to which is added Notes on the Principal Manufacturing Centres and Remarkable Manufactories. By Edward Young and Edwin T. Freedley. In two vols., 8vo. Vol. 1 now ready. Price,..............................$3.00

Bookbinding: A Manual of the Art of Bookbinding,

Containing full instructions in the different branches of Forwarding, Gilding and Finishing. Also, the Art of Marbling Book-edges and Paper. By James B. Nicholson. Illustrated. 12mo., cloth,..............................$1.75

CONTENTS.—Sketch of the Progress of Bookbinding, Sheet-work, Forwarding the Edges, Marbling, Gilding the Edges, Covering, Half Binding, Blank Binding, Boarding, Cloth-work, Ornamental Art, Finishing, Taste and Design, Styles, Gilding, Illuminated Binding, Blind Tooling, Antique, Coloring, Marbling, Uniform Colors, Gold Marbling, Landscapes, etc., Inlaid Ornaments, Harmony of Colors, Pasting Down, etc., Stamp or Press-work, Restoring the Bindings of Old Books, Supplying imperfections in Old Books, Hints to Book Collectors, Technical Lessons.

Booth and Morfit. The Encyclopedia of Chemistry, Practical and Theoretical:

Embracing its application to the Arts, Metallurgy, Mineralogy, Geology, Medicine, and Pharmacy, By JAMES C. BOOTH, Melter and Refiner in the United States Mint; Professor of Applied Chemistry in the Franklin Institute, etc.; assisted by CAMPBELL MORFIT, author of "Chemical Manipulations," etc. 7th Edition. Complete in one volume, royal octavo, 978 pages, with numerous wood cuts and other illustrations,..............................$5.00

From the very large number of articles in this volume, it is entirely impossible to give a list of the Contents, but attention may be called to some among the more elaborate, such as Affinity, Alcoholometry, Ammonium, Analysis, Antimony, Arsenic, Blowpipes, Cyanogen, Distillation, Electricity, Ethyl, Fermentation, Iron, Lead and Water.

Brewer; (The Complete Practical.)

Or Plain, Concise, and Accurate Instructions in the Art of Brewing Beer, Ale, Porter, etc., etc., and the Process of Making all the Small Beers. By M. LAFAYETTE BYRN, M. D. With Illustrations. 12mo..............................$1.00

"Many an old brewer will find in this book valuable hints and sug-

PUBLISHED BY HENRY CAREY BAIRD.

gestions worthy of consideration, and the novice can post himself up in his trade in all its parts."—*Artisan.*

Builder's Pocket Companion:

Containing the Elements of Building, Surveying, and Architecture; with Practical Rules and Instructions connected with the subject. By A. C. SMEATON, Civil Engineer, etc. In one volume, 12mo., ...$1.00

CONTENTS.—The Builder, Carpenter, Joiner, Mason, Plasterer, Plumber, Painter, Smith, Practical Geometry, Surveyor, Cohesive Strength of Bodies, Architect.

"It gives, in a small space, the most thorough directions to the builder, from the laying of a brick, or the felling of a tree, up to the most elaborate production of ornamental architecture. It is scientific, without being obscure and unintelligible; and every house-carpenter, master, journeyman, or apprentice, should have a copy at hand always."—*Evening Bulletin.*

Byrne. The Handbook for the Artisan, Mechanic, and Engineer,

Containing Instructions in Grinding and Sharpening of Cutting Tools, Figuration of Materials by Abrasion, Lapidary Work, Gem and Glass Engraving, Varnishing and Lackering, Abrasive Processes, etc., etc. By Oliver Byrne. Illustrated with 11 large plates and 185 cuts. 8vo., cloth,..$5.00

CONTENTS.—Grinding Cutting Tools on the Ordinary Grindstone; Sharpening Cutting Tools on the Oilstone; Setting Razors; Sharpening Cutting Tools with Artificial Grinders; Production of Plane Surfaces by Abrasion; Production of Cylindrical Surfaces by Abrasion; Production of Conical Surfaces by Abrasion; Production of Spherical Surfaces by Abrasion; Glass Cutting; Lapidary Work; Setting, Cutting, and Polishing Flat and Rounded Works; Cutting Faucets; Lapidary Apparatus for Amateurs; Gem and Glass Engraving; Seal and Gem Engraving; Cameo Cutting; Glass Engraving, Varnishing, and Lackering; General Remarks upon Abrasive Processes; Dictionary of Apparatus; Materials and Processes for Grinding and Polishing commonly employed in the Mechanical and Useful Arts.

Byrne. The Practical Metal-worker's Assistant,

For Tin-plate Workers, Braziers, Coppersmiths, Zincplate Ornamenters and Workers, Wire Workers, Whitesmiths, Blacksmiths, Bell Hangers, Jewellers, Silver and Gold Smiths, Electrotypers, and all other Workers in Alloys and Metals. Edited by OLIVER BYRNE. Complete in one volume, octavo, ...$7.50

It treats of Casting, Founding, and Forging; of Tongs and other Tools; Degrees of Heat and Management of Fires; Welding of

PRACTICAL AND SCIENTIFIC BOOKS,

Heading and Swage Tools; of Punches and Anvils; of Hardening and Tempering; of Malleable Iron Castings, Case Hardening, Wrought and Cast Iron; the Management and Manipulation of Metals and Alloys, Melting and Mixing; the Management of Furnaces, Casting and Founding with Metallic Moulds, Joining and Working Sheet Metal; Peculiarities of the different Tools employed; Processes dependent on the ductility of Metals; Wire Drawing, Drawing Metal Tubes, Soldering; The use of the Blowpipe, and every other known Metal Worker's Tool.

Byrne. The Practical Model Calculator,

For the Engineer, Machinist, Manufacturer of Engine Work, Naval Architect, Miner, and Millwright. By OLIVER BYRNE, Compiler and Editor of the Dictionary of Machines, Mechanics, Engine Work and Engineering, and Author of various Mathematical and Mechanical Works. Illustrated by numerous engravings. Complete in one large volume, octavo, of nearly six hundred pages,..$3.50

The principal objects of this work are: to establish model calculations to guide practical men and students; to illustrate every practical rule and principle by numerical calculations, systematically arranged; to give information and data indispensable to those for whom it is intended, thus surpassing in value any other book of its character; to economize the labor of the practical man, and to render his every-day calculations easy and comprehensive. It will be found to be one of the most complete and valuable practical books ever published.

Cabinetmaker's and Upholsterer's Companion,

Comprising the Rudiments and Principles of Cabinetmaking and Upholstery, with Familiar Instructions, illustrated by Examples for attaining a proficiency in the Art of Drawing, as applicable to Cabinet Work; the processes of Veneering, Inlaying, and Buhl Work; the Art of Dyeing and Staining Wood, Bone, Tortoise Shell, etc. Directions for Lackering, Japanning, and Varnishing; to make French Polish; to prepare the best Glues, Cements, and Compositions, and a number of Receipts particularly useful for Workmen generally. By J. STOKES. In one volume, 12mo. With Illustrations,............... 75

"A large amount of practical information, of great service to all concerned in those branches of business."—*Ohio State Journal.*

Campion. A Practical Treatise on Mechanical Engineering;

Comprising Metallurgy, Moulding, Casting, Forging Tools, Workshop Machinery, Mechanical Manipulation, Manufacture of Steam Engine, etc., etc. Illustrated with 28 plates of Boilers, Steam Engines, Workshop Machinery, etc., and 91 Wood Engravings; with an Appendix on the Analysis of Iron and Iron Ores. By Francis Campion, C. E., President of the Civil and Mechanical Engineers' Society, etc. (*In press.*)

Celnart. The Perfumer.

From the French of Madame Celnart; with additions by Professor H. Dussauce. 8vo. (*In press.*)

Colburn. The Locomotive Engine;

Including a Description of its Structure, Rules for Estimating its Capabilities, and Practical Observations on its Construction and Management. By ZERAH COLBURN. Illustrated. A new edition. 12mo,............................ 75

"It is the most practical and generally useful work on the Steam Engine that we have seen."—*Boston Traveler.*

Daguerreotypist and Photographer's Companion.

12mo., cloth,..$1.00

Distiller (The Complete Practical).

By M. LAFAYETTE BYRN, M.D. With Illustrations. 12mo. $1.00

"So simplified, that it is adapted not only to the use of extensive Distillers, but for every farmer, or others who may want to engage in Distilling."—*Banner of the Union.*

Dussauce. Practical Treatise

ON THE FABRICATION OF MATCHES, GUN COTTON, AND FULMINATING POWDERS. By Prof. H. Dussauce. (*In press.*)

CONTENTS.—*Phosphorus.*—History of Phosphorus; Physical Properties; Chemical Properties; Natural State; Preparation of White Phosphorus; Amorphous Phosphorus, and Benoxide of Lead. *Matches.*—Preparation of Wooden Matches; Matches inflammable by rubbing, without noise; Common Lucifer Matches: Matches without Phosphorus; Candle Matches; Matches with Amorphous Phosphorus; Matches and Rubbers without Phosphorus. *Gun Cotton.*—Properties; Preparation; Paper Powder; use of Cotton and Paper Powders for Fulminating Primers, etc.; Preparation of Fulminating Primers, etc., etc.

Dussauce. Chemical Receipt Book:

A General Formulary for the Fabrication of Leading Chemicals, and their Application to the Arts, Manufactures, Metallurgy, and Agriculture. By Prof. H. Dussauce. (*In press.*)

DYEING, CALICO PRINTING, COLORS, COTTON SPINNING, AND WOOLEN MANUFACTURE.

Baird. The American Cotton Spinner, and Manager's and Carder's Guide:

A Practical Treatise on Cotton Spinning; giving the Dimensions and Speed of Machinery, Draught and Twist Calculations, etc.; with Notices of recent Improvements: together with Rules and Examples for making changes in the sizes and numbers of Roving and Yarn. Compiled from the papers of the late Robert H. Baird. 12mo..........$1.25

Capron De Dole. Dussauce. Blues and Carmines of Indigo:

A Practical Treatise on the Fabrication of every Commercial Product derived from Indigo. By Felicien Capron de Dole. Translated, with important additions, by Professor H. Dussauce. 12mo..........$2.50

Chemistry Applied to Dyeing.

By James Napier, F. C. S. Illustrated. 12mo........$2.00

CONTENTS.—*General Properties of Matter.*—Heat, Light, Elements of Matter, Chemical Affinity. *Non-Metallic Substances.*—Oxygen, Hydrogen, Nitrogen, Chlorine, Sulphur, Selenium, Phosphorus, Iodine, Bromine, Fluorine, Silicum, Boron, Carbon. *Metallic Substances.*—General Properties of Metals, Potassium, Sodium, Lithium, Soap, Barium. Strontium, Calcium, Magnesium, Alminum, Manganese, Iron, Cobalt, Nickel, Zinc, Cadmium, Copper, Lead, Bismuth, Tin, Titanium, Chromium, Vanadium, Tungstenum or Wolfram, Molybdenum, Tellarium, Arsenic, Antimony, Uranium, Cerium, Mercury, Silver, Gold, Platinum, Palladium, Iridium, Osmium, Rhodium, Lanthanium. *Mordants.*—Red Spirits, Barwood Spirits, Plumb Spirits, Yellow Spirits, Nitrate of Iron, Acetate of Alumina, Black Iron Liquor, Iron and Tin for Royal Blues, Acetate of Copper. *Vegetable Matters used in Dyeing.*—Galls, Sumach, Catechu, Indigo, Logwood, Brazil-woods, Sandal-wood, Barwood, Camwood, Fustic, Young Fustic, Bark or Quercitron, Flavine, Weld or Wold, Turmeric, Persian Berries, Safflower, Madder, Munjeet, Annota, Alkanet Root, Archil. *Proposed New Vegetable Dyes.*—Sooranjee, Carajuru, Wongshy, Aloes, Pittacal, Barbary Root. *Animal Matters used in Dyeing.*—Cochineal, Lake or Lac, Kerms.

This will be found one of the most valuable books on the subject of dyeing, ever published in this country.

Dussauce. Treatise on the Coloring Matters Derived from Coal Tar;

Their Practical Application in Dyeing Cotton, Wool, and

PUBLISHED BY HENRY CAREY BAIRD.

Silk; the Principles of the Art of Dyeing and of the Distillation of Coal Tar; with a Description of the most Important New Dyes now in use. By Professor H. Dussauce, Chemist. 12mo............................$2.50

CONTENTS.—Historical Notice of the Art of Dyeing—Chemical Principles of the Art of Dyeing—Preliminary Preparation of Stuffs—Mordants—Dyeing—On the Coloring Matters produced by Coal Tar—Distillation of Coal Tar—History of Aniline—Properties of Aniline—Preparation of Aniline directly from Coal Tar—Artificial Preparation of Aniline—Preparation of Benzole—Properties of Benzole—Preparation of Nitro-Benzole—Transformation of Nitro-Benzole into Aniline, by means of Sulphide of Ammonium; by Nascent Hydrogen; by Acctate of Iron; and by Arsenite of Potash—Properties of the Bi-Nitro-Benzole—Aniline Purple—Violine—Roseine—Emeraldine—Bleu de Paris—Futschine, or Magenta—Coloring Matters obtained by other bases from Coal Tar—Nitroso-Phenyline—Di Nitro-Aniline—Nitro-Phenyline—Picric Acid—Rosolic Acid—Quinoline—Napthaline Colors—Chloroxynaphthalic and Perchloroxynapthalic Acids—Carminaphtha—Ninaphthalamine—Nitrosonaphthaline—Naphthamein—Tar Red—Azuline—Application of Coal Tar Colors to the Art of Dyeing and Calico Printing—Action of Light on Coloring Matters from Coal Tar—Latest Improvements in the Art of Dyeing—Chrysammic Acid—Molybdic and Picric Acids—Extract of Madder—Theory of the Fixation of Coloring Matters in Dyeing and Printing—Principles of the Action of the most important Mordants—Aluminous Mordants—Ferruginous Mordants—Stanniferous Mordants—Artificial Alizarin—Metallic Hyposulphites as Mordants—Dyer's Soap—Preparation of Indigo for Dyeing and Printing—Relative Value of Indigo—Chinese Green Murexide.

Dyer and Color-maker's Companion:

Containing upwards of two hundred Receipts for making Colors, on the most approved principles, for all the various styles and fabrics now in existence; with the Scouring Process, and plain Directions for Preparing, Washing-off, and Finishing the Goods. Second edition. In one volume, 12mo... 75

French Dyer, (The):

Comprising the Art of Dyeing in Woolen, Silk, Cotton, etc., etc. By M. M. Riffault, Vernaud, De Fonteuelle, Thillaye, and Mallepeyre. (*In press.*)

Love. The Art of Dyeing, Cleaning, Scouring, and Finishing,

ON THE MOST APPROVED ENGLISH AND FRENCH METHODS; being Practical Instructions in Dyeing Silks, Woolens and Cottons, Feathers, Chips, Straw, etc., Scouring and Cleaning Bed and Window Curtains, Carpets, Rugs, etc., French and English Cleaning, any Color or Fabric of Silk, Satin, or Damask. By Thomas Love, a working Dyer and Scourer. In one volume, 12mo............$3.00

O'Neill. Chemistry of Calico Printing, Dyeing, and Bleaching;

Including Silken, Woolen, and Mixed Goods; Practical and Theoretical. By Charles O'Neill. (*In press.*)

O'Neill. A Dictionary of Calico Printing and Dyeing.

By Charles O'Neill. (*In press.*)

Scott. The Practical Cotton-spinner and Manufacturer;

OR, THE MANAGER AND OVERLOOKER'S COMPANION. This work contains a Comprehensive System of Calculations for Mill Gearing and Machinery, from the first Moving Power, through the different processes of Carding, Drawing, Slabbing, Roving, Spinning, and Weaving, adapted to American Machinery, Practice and Usages. Compendious Tables of Yarns and Reeds are added. Illustrated by large Working-Drawings of the most approved American Cotton Machinery. Complete in one volume, octavo... $3.50

This edition of Scott's Cotton-Spinner, by Oliver Byrne, is designed for the American Operative. It will be found intensely practical, and will be of the greatest possible value to the Manager, Overseer, and Workman.

Sellers. The Color-mixer.

By John Sellers, an Experienced Practical Workman. To which is added a CATECHISM OF CHEMISTRY. In one volume, 12mo. (*In press.*)

Smith. The Dyer's Instructor;

Comprising Practical Instructions in the Art of Dyeing Silk, Cotton, Wool and Worsted, and Woolen Goods, as Single and Two-colored Damasks, Moreens, Camlets, Lastings, Shot Cobourgs, Silk Striped Orleans, Plain Orleans, from White and Colored Warps, Merinos, Woolens, Yarns, etc.; containing nearly eight hundred Receipts. To which is added a Treatise on the Art of Padding, and the Printing of Silk Warps, Skeins and Handkerchiefs, and the various Mordants and Colors for the different styles of such work. By David Smith, Pattern Dyer. A new edition, in one volume, 12mo......................$3.00

CONTENTS.—Wool Dyeing, 60 receipts—Cotton Dyeing, 68 receipts—Silk Dyeing, 60 receipts—Woolen Yarn Dyeing, 59 receipts—Worsted Yarn Dyeing, 51 receipts—Woolen Dyeing, 52 receipts—Damask Dyeing, 40 receipts—Moreen Dyeing, 38 receipts—Two-Colored Damask Dyeing, 21 receipts—Camlet Dyeing, 23 receipts—Lasting Dyeing, 23 receipts—Shot Cobourg Dyeing, 18 receipts—Silk Striped Orleans, from Black, White, and Colored Warps, 23 receipts—Colored Orleans, from Black Warps, 15 receipts—Colored Orleans and Cobourgs, from White Warps, 27 receipts—Colored Merinos, 41 receipts—Woolen Shawl Dyeing, 15 receipts—Padding, 42 receipts—Silk Warp, Skein, and Handkerchief Printing, 62 receipts—Nature and Use of Dyewares, including Alum, Annotta, Archil, Ammonia, Argol, Super Argol, Camwood, Catechu, Cochineal, Chrome, or Bichromate of Potash, Cudbear, Chemic, or Sulphate of Indigo, French Berry, or Persian Berry, Fustic or Young Fustic, Galls, Indigo, Kermes or Lac Dye, Logwood, Madder, Nitric Acid or Aqua Fortis, Nitrates, Oxalic Tin, Peachwood, Prussiate of Potash, Quercitron Bark, Safflower, Saunders or Red Sandal, Sapan Wood, Sumach, Turmeric, Examination of Water by Tests, etc., etc.

Toustain. A Practical Treatise on the Woolen Manufacture.

From the French of M. Toustain. (*In press.*)

Ulrich. Dussauce. A Complete Treatise

ON THE ART OF DYEING COTTON AND WOOL, AS PRACTISED IN PARIS, ROUEN, MULHOUSE AND GERMANY. From the French of M. Louis Ulrich, a Practical Dyer in the principal Manufactories of Paris, Rouen, Mulhouse, etc., etc.; to which are added the most important Receipts for Dyeing Wool, as practised in the Manufacture Imperiale des Gobelins, Paris. By Professor H. Dussauce. 12mo..$3.00

CONTENTS.—
Rouen Dyes, 106 Receipts.
Alsace   "   235   "
German   "   109   "
Mulhouse "   72   "
Parisian "   56   "
Gobelins "   100   "
In all nearly 700 Receipts.

Easton. A Practical Treatise on Street or Horse-power Railways;

Their Location, Construction and Management; with general Plans and Rules for their Organization and Operation; together with Examinations as to their Comparative Advantages over the Omnibus System, and Inquiries as to their Value for Investment; including Copies of Municipal Ordinances relating thereto. By Alexander Easton, C. E. Illustrated by twenty-three plates, 8vo., cloth..$2.00

Examinations of Drugs, Medicines, Chemicals, etc.,

As to their Purity and Adulterations. By C. H. Peirce, M. D. 12mo., cloth...$2.00

Fisher's Photogenic Manipulation.

16mo., cloth.. 62

Gas and Ventilation;

A Practical Treatise on Gas and Ventilation. By E. E. Perkins. 12mo., cloth.. 75

Gilbart. A Practical Treatise on Banking.

By James William Gilbart, F. R. S. A new enlarged and improved edition. Edited by J. Smith Homans, editor of "Banker's Magazine." To which is added "Money," by H. C. Carey. 8vo..$3.00

Gregory's Mathematics for Practical Men;

Adapted to the Pursuits of Surveyors, Architects, Mechanics and Civil Engineers. 8vo., plates, cloth...$1.50

Hardwich. A Manual of Photographic Chemistry;

Including the practice of the Collodion Process. By J. F. Hardwich. (*In press.*)

Hay. The Interior Decorator;

The Laws of Harmonious Coloring adapted to Interior Decorations; with a Practical Treatise on House Painting. By D. R. Hay, House Painter and Decorator. Illustrated by a Diagram of the Primary, Secondary and Tertiary Colors. 12mo. (*In press.*)

PUBLISHED BY HENRY CAREY BAIRD.

Inventor's Guide—Patent Office and Patent Laws:

Or, a Guide to Inventors, and a Book of Reference for Judges, Lawyers, Magistrates, and others. By J. G. Moore. 12mo., cloth.................................$1.00

Jervis. Railway Property. A Treatise

ON THE CONSTRUCTION AND MANAGEMENT OF RAILWAYS; designed to afford useful knowledge, in the popular style, to the holders of this class of property; as well as Railway Managers, Officers and Agents. By John B. Jervis, late Chief Engineer of the Hudson River Railroad, Croton Aqueduct, etc. One volume, 12mo., cloth........$1.50

CONTENTS.— Preface—Introduction. *Construction.*— Introductory—Land and Land Damages—Location of Line—Method of Business—Grading—Bridges and Culverts—Road Crossings—Ballasting Track—Cross Sleepers—Chairs and Spikes—Rails—Station Buildings—Locomotives, Coaches and Cars. *Operating.*—Introductory—Freight—Passengers—Engine Drivers—Repairs to Track—Repairs of Machinery—Civil Engineer—Superintendent—Supplies of Material—Receipts—Disbursements—Statistics—Running Trains—Competition—Financial Management—General Remarks.

Johnson. The Coal Trade of British America;

With Researches on the Characters and Practical Values of American and Foreign Coals. By Walter R. Johnson, Civil and Mining Engineer and Chemist. 8vo........$2.00

This volume contains the results of the experiments made for the Navy Department, upon which their Coal contracts are now based.

Johnston. Instructions for the Analysis of Soils, Limestones and Manures.

By J. F. W. Johnston. 12mo........................ 38

Larkin. The Practical Brass and Iron Founder's Guide;

A Concise Treatise on the Art of Brass Founding, Moulding, etc. By James Larkin. 12mo., cloth............$1.00

Leslie's (Miss) Complete Cookery;

Directions for Cookery in its Various Branches. By Miss Leslie. 58th thousand. Thoroughly revised; with the addition of New Receipts. In one volume, 12mo., half bound, or in sheep..$1.00

13

Leslie's (Miss) Ladies' House Book;
A Manual of Domestic Economy. 20th revised edition. 12mo., sheep $1.00

Leslie's (Miss) Two Hundred Receipts in French Cookery.
Cloth, 12mo.. 25

Lieber. Assayer's Guide;
Or, Practical Directions to Assayers, Miners and Smelters, for the Tests and Assays, by Heat and by Wet Processes, of the Ores of all the principal Metals; and of Gold and Silver Coins and Alloys. By Oscar M. Lieber, late Geologist to the State of Mississippi. 12mo. With illustrations 75

"Among the indispensable works for this purpose, is this little guide."—*Artizan*.

Lowig. Principles of Organic and Physiological Chemistry.
By Dr. Carl Löwig, Doctor of Medicine and Philosophy; Ordinary Professor of Chemistry in the University of Zürich; Author of "Chemie des Organischen Verbindungen." Translated by Daniel Breed, M. D., of the U. S. Patent Office; late of the Laboratory of Liebig and Lowig. 8vo., sheep..$3.50

Marble Worker's Manual;
Containing Practical Information respecting Marbles in general, their Cutting, Working and Polishing, Veneering, etc., etc. 12mo., cloth......................................$1.00

Miles. A Plain Treatise on Horse-shoeing.
With Illustrations. By William Miles, Author of "The Horse's Foot.".. 75

Morfit. The Arts of Tanning, Currying and Leather Dressing.
Theoretically and Practically Considered in all their Details; being a Full and Comprehensive Treatise on the

PUBLISHED BY HENRY CAREY BAIRD.

Manufacture of the Various Kinds of Leather. Illustrated by over two hundred Engravings. Edited from the French of De Fontenelle and Malapeyre. With numerous Emendations and Additions, by Campbell Morfit, Practical and Analytical Chemist. Complete in one volume, octavo.. ....$10.00

This important Treatise will be found to cover the whole field in the most masterly manner, and it is believed that in no other branch of applied science could more signal service be rendered to American Manufactures.

The publisher is not aware that in any other work heretofore issued in this country, more space has been devoted to this subject than a single chapter; and in offering this volume to so large and intelligent a class as American Tanners and Leather Dressers, he feels confident of their substantial support and encouragement.

CONTENTS.—Introduction—Dignity of Labor—Tan and Tannin—Gallic Acid—Extractive-Tanning Materials—Oak Barks—Barking of Trees—Method of Estimating the Tanning Power of Astringent Substances—Tan—The Structure and Composition of Skin—Different Kinds of Skin suitable for Tanning—Preliminary Treatment of Skins—Tanning Process—Improved Processes—Vauquelin's Process—Accelerating Processes—Keasley's, Trumbull's, Hibbard's, and Leprieur's Processes—Tanning with Extract of Oak-Bark—Hemlock Tanning—With Myrtle Plant—English Harness Leather—Calf Skins—Goat and Sheep Skins—Horse Hides—Buck, Wolf and Dog Skins—Buffalo, or "Grecian" Leather—Russia Leather—Red Skins—Wallachia Leather—Mineral Tanning—Texture and Quality of Leather, and the means of Discovering its Defects—Tawing—Hungary Leather—Oiled Leather—Tanning as practised by the Mongol Tartars—Shagreen—Parchment—Leather Bottles—Tanning of Cordage and Sail Cloth—Glazed or "Patent" Leather—Helverson's Process for Rendering Hides Hard and Transparent—Currying—Currying of Calf Skins—Currying of Goat Skins—Red Leather—Fair Leather—Water Proof Dressing—Perkins' Machine for Pommelling and Graining Leather—Splitting, Shaving, Fleshing and Cleansing Machines—Embossing of Leather—Gut Dressing..

Morfit. A Treatise on Chemistry

APPLIED TO THE MANUFACTURE OF SOAP AND CANDLES; being a Thorough Exposition, in all their Minutiæ, of the principles and Practice of the Trade, based upon the most recent Discoveries in Science and Art. By Campbell Morfit, Professor of Analytical and Applied Chemistry in the University of Maryland. A new and improved edition. Illustrated with 260 Engravings on Wood. Complete in one volume, large 8vo............................$6.00

CONTENTS.—CHAPTER I. The History of the Art and its Relations to Science—II. Chemical Combination—III. Alkalies and Alkaline Earths—IV. Alkalimentary—V. Acids—VI. Origin and Composition of Fatty Matters—VII. Saponifiable Fats—Vegetable Fats—Animal Fats—Waxes—VIII. Action of Heat and Mineral Acids of Fatty Matters—IX. Volatile or Essential Oils, and Resins—X. The Proximate Principles of Fats—Their Composition and Properties—Basic Constituents of Fats—XI. Theory of Saponification—XII. Utensils Requisite for a Soap Factory—XIII. Preparatory Manipulations in the Process of Making Soap—Preparation of the Lyes—XIV. Hard

Soaps—XV. Soft Soaps—XVI. Soaps by the Cold Process—XVII. Silicated Soaps—XVIII. Toilet Soaps—XIX. Patent Soaps—XX. Fraud and Adulterations in the Manufacture of Soap—XXI. Candles—XXII. Illumination—XXIII. Philosophy of Flame—XXIV. Raw Material for Candles—Purification and Bleaching of Suet—XXV. Wicks—XXVI. Dipped Candles—XXVII. Moulded Candles—XXVIII. Stearin Candles—XXIX. Stearic Acid Candles—"Star" or "Adamantine" Candles—Saponification by Lime—Saponification by Lime and Sulphurous Acid—Saponification by Sulphuric Acid—Saponification by the combined action of Heat, Pressure and Steam—XXX. Spermaceti Candles—XXXI. Wax Candles—XXXII. Composite Candles—XXXIII. Paraffin—XXXIV. Patent Candles—XXXV. Hydrometers and Thermometers.

Mortimer. Pyrotechnist's Companion;

Or, a Familiar System of Fire-works. By G. W. Mortimer. Illustrated by numerous Engravings. 12mo... 75

Napier. Manual of Electro-Metallurgy;

Including the Application of the Art to Manufacturing Processes. By James Napier. From the second London edition, revised and enlarged. Illustrated by Engravings. In one volume, 12mo................$1.50

Napier's Electro-Metallurgy is generally regarded as the very best Practical Treatise on the Subject in the English Language.

CONTENTS.—History of the Art of Electro-Metallurgy—Description of Galvanic Batteries, and their respective Peculiarities—Electrotype Processes—Miscellaneous Applications of the Process of Coating with Copper—Bronzing—Decomposition of Metals upon one another—Electro-Plating—Electro-Gilding—Results of Experiments on the Deposition of other Metals as Coatings, Theoretical Observations.

Norris's Hand-book for Locomotive Engineers and Machinists;

Comprising the Calculations for Constructing Locomotives, Manner of setting Valves, etc., etc. By Septimus Norris, Civil and Mechanical Engineer. In one volume, 12mo., with Illustrations................$1.50

"With pleasure do we meet with such a work as Messrs. Norris and Baird have given us."—*Artizan.*

"In this work he has given us what are called 'the secrets of the business,' in the rules to construct locomotives, in order that the million should be learned in all things."—*Scientific American.*

Nystrom. A Treatise on Screw-Propellers and their Steam-Engines;

With Practical Rules and Examples by which to Calculate and Construct the same for any description of Vessels. By J. W. Nystrom. Illustrated by over thirty large Working Drawings. In one volume, octavo...$3.50

PUBLISHED BY HENRY CAREY BAIRD.

Overman. The Manufacture of Iron in all its Various Branches;

To which is added an Essay on the Manufacture of Steel. By Frederick Overman, Mining Engineer. With one hundred and fifty Wood Engravings. Third edition. In one volume, octavo, five hundred pages................ $6.00

"We have now to announce the appearance of another valuable work on the subject, which, in our humble opinion, supplies any deficiency which late improvements and discoveries may have caused, from the lapse of time since the date of 'Mushet' and 'Schrivenor.' It is the production of one of our Trans-Atlantic brethren, Mr. Frederick Overman, Mining Engineer; and we do not hesitate to set it down as a work of great importance to all connected with the iron interests; one which, while it is sufficiently technological fully to explain chemical analysis, and the various phenomena of iron under different circumstances, to the satisfaction of the most fastidious, is written in that clear and comprehensive style as to be available to the capacity of the humblest mind, and consequently will be of much advantage to those works where the proprietors may see the desirability of placing it in the hands of their operatives."—*London Mining Journal.*

Painter, Gilder and Varnisher's Companion;

Containing Rules and Regulations in every thing relating to the Arts of Painting, Gilding, Varnishing and Glass Staining; with numerous useful and valuable Receipts; Tests for the detection of Adulterations in Oils and Colors; and a statement of the Diseases and Accidents to which Painters, Gilders and Varnishers are particularly liable, with the simplest methods of Prevention and Remedy. Eighth edition. To which are added Complete Instructions in Graining, Marbling, Sign Writing, and Gilding on Glass. 12mo., cloth............................ 75

Paper-Hanger's (The) Companion;

In which the Practical Operations of the Trade are systematically laid down; with copious Directions Preparatory to Papering; Preventions against the effect of Damp in Walls; the various Cements and Pastes adapted to the several purposes of the Trade; Observations and Directions for the Panelling and Ornamenting of Rooms, etc., etc. By James Arrowsmith. In one volume, 12mo 75

Practical (The) Surveyor's Guide;

Containing the necessary information to make any person of common capacity a finished Land Surveyor, without the aid of a Teacher. By Andrew Duncan, Land Surveyor and Civil Engineer. 12mo.................... 75

Having had an experience as a Practical Surveyor, etc., of thirty years, it is believed that the author of this volume possesses a thorough knowledge of the wants of the profession; and never having met with any work sufficiently concise and instructive in the several details necessary for the proper qualification of the Surveyor, it has been his object to supply that want. Among other important matters in the book, will be found the following:

Instructions in levelling and profiling, with a new and speedy plan of setting grades on rail and plank roads—the method of inflecting curves—the description and design of a new instrument, whereby distances are found at once, without any calculation—a new method of surveying any tract of land by measuring one line through it—a geometrical method of correcting surveys taken with the compass, to fit them for calculation—a short method of finding the angles from the courses, and *vice versa*—the method of surveying with the compass through any mine or iron works, and to correct the deflections of the needle by attraction—description of an instrument by the help of which any one may measure a map by inspection, without calculation—a new and short method of calculation, wherein fewer figures are used—the method of correcting the diurnal variation of the needle—various methods of plotting and embellishing maps—the most correct method of laying off plots with the pole, etc.—description of a new compass contrived by the author, etc., etc.

Railroad Engineer's Pocket Companion for the Field.

By W. Griswold. 12mo., tucks.....................$1.00

Riddell. The Elements of Hand-Railing;

Being the most Complete and Original Exposition of this Branch of Carpentry that has appeared. By Robert Riddell. Third edition. Enlarged and improved. Illustrated by 22 large plates. 4to., cloth..............$3.00

Rural Chemistry;

An Elementary Introduction to the Study of the Science, in its relation to Agriculture and the Arts of Life. By Edward Solly, Professor of Chemistry in the Horticultural Society of London. From the third improved London edition. 12mo...................................$1.25

Shunk. A Practical Treatise

ON RAILWAY CURVES, AND LOCATION FOR YOUNG ENGINEERS. By Wm. F. Shunk, Civil Engineer. 12mo............$1.00

Strength and Other Properties of Metals;

Reports of Experiments on the Strength and other Properties of Metals for Cannon. With a Description of the Machines for Testing Metals, and of the Classification of Cannon in service. By Officers of the Ordnance Department U. S. Army. By authority of the Secretary of War. Illustrated by 25 large steel plates. In one volume, quarto..$10.00

The best Treatise on Cast-iron extant.

Tables Showing the Weight

OF ROUND, SQUARE AND FLAT BAR IRON, STEEL, etc., by Measurement. Cloth... 50

Taylor. Statistics of Coal;

Including Mineral Bituminous Substances employed in Arts and Manufactures; with their Geographical, Geological and Commercial Distribution, and Amount of Production and Consumption on the American Continent. With Incidental Statistics of the Iron Manufacture. By R. C. Taylor. Second edition, revised by S. S. Haldeman. Illustrated by five Maps and many Wood Engravings. 8vo., cloth..$6.00

Templeton. The Practical Examiner on Steam and the Steam Engine;

With Instructive References relative thereto, arranged for the use of Engineers, Students, and others. By Wm. Templeton, Engineer. 12mo............................... 75

This work was originally written for the author's private use. He was prevailed upon by various Engineers, who had seen the notes, to consent to its publication, from their eager expression of belief that it would be equally useful to them as it had been to himself.

Tin and Sheet Iron Worker's Instructor;

Comprising complete Descriptions of the necessary Patterns and Machinery, and the Processes of Calculating Dimensions, Cutting, Joining, Raising, Soldering, etc., etc. With numerous Illustrations. (*In press.*)

Treatise (A) on a Box of Instruments,

And the Slide Rule; with the Theory of Trigonometry and Logarithms, including Practical Geometry, Surveying, Measuring of Timber, Cask and Malt Gauging,

PRACTICAL AND SCIENTIFIC BOOKS,

Heights and Distances. By Thomas Kentish. In one volume, 12mo...$1.00

A volume of inestimable value to Engineers, Gaugers, Students, and others.

Turnbull. The Electro-Magnetic Telegraph;

With an Historical Account of its Rise, Progress, and Present Condition. Also, Practical Suggestions in regard to Insulation and Protection from the Effects of Lightning. Together with an Appendix containing several important Telegraphic Devices and Laws. By Lawrence Turnbull, M. D., Lecturer on Technical Chemistry at the Franklin Institute. Second edition. Revised and improved. Illustrated by numerous Engravings. 8vo..$2.00

Turner's (The) Companion;

Containing Instruction in Concentric, Elliptic and Eccentric Turning; also various Steel Plates of Chucks, Tools and Instruments; and Directions for Using the Eccentric Cutter, Drill, Vertical Cutter and Rest; with Patterns and Instructions for working them. 12mo., cloth..... 75

Bell. Carpentry Made Easy;

Or, The Science and Art of Framing, on a New and Improved System; with Specific Instructions for Building Balloon Frames, Barn Frames, Mill Frames, Warehouses, Church Spires, etc.; comprising also a System of Bridge Building; with Bills, Estimates of Cost, and Valuable Tables. Illustrated by 38 plates, comprising nearly 200 figures. By William E. Bell, Architect and Practical Builder. 8vo...$3.60

SOCIAL SCIENCE.
THE WORKS OF HENRY C. CAREY.

"I challenge the production from among the writers on political economy of a more learned, philosophical, and convincing speculator on that theme, than my distinguished fellow-citizen, Henry C. Carey. The works he has published in support of the protective policy, are remarkable for profound research, extensive range of inquiry, rare logical acumen, and a consummate knowledge of history."—*Speech of Hon. Edward Joy Morris, in the House of Representatives of the United States, February 2, 1859.*

PUBLISHED BY HENRY CAREY BAIRD.

THE WORKS OF HENRY C. CAREY.

"Henry C. Carey, the best known and ablest economist of North America. * * * * * In Europe he is principally known by his striking and original attacks, based upon the peculiar advantages of American experience, on some of the principal doctrines, especially Malthus' 'Theory of Population' and Ricardo's teachings. His views have been largely adopted and thoroughly discussed in Europe."—*"The German Political Lexicon," Edited by Bluntschli and Brater. Leipsic, 1858.*

"We believe that your labors mark an era in the science of political economy. To your researches and lucid arguments are we indebted for the explosion of the absurdities of Malthus, Say, and Ricardo, in regard to the inability of the earth to meet the demands of a growing population. American industry owes you a debt which cannot be repaid, and which it will ever be proud to acknowledge.—*From a Letter of Hon. George W. Scranton, M. C., Hon. William Jessup, and over sixty influential citizens of Luzerne County, Pennsylvania, to Henry C. Carey, April 3, 1859.*

Financial Crises;
Their Causes and Effects. 8vo., paper...... 25

French and American Tariffs,
Compared in a Series of Letters addressed to Mons. M. Chevalier. 8vo., paper...... 15

Harmony (The) of Interests;
Agricultural, Manufacturing and Commercial. 8vo., paper...... 75
Cloth...... $1.25

"We can safely recommend this remarkable work to all who wish to investigate the causes of the progress or decline of industrial communities."—*Blackwood's Magazine.*

Letters to the President of the United States.
8vo., Paper...... 50

Miscellaneous Works;
Comprising "Harmony of Interests," "Money," "Letters to the President," "French and American Tariffs," and "Financial Crises." One volume, 8vo., half bound. $2.25

Money; A Lecture
Before the New York Geographical and Statistical Society. 8vo., paper...... 15

PRACTICAL AND SCIENTIFIC BOOKS.

THE WORKS OF HENRY C. CAREY.

Past (The), the Present, and the Future.

8vo..$2.00
12mo...$1.25

"Full of important facts bearing on topics that are now agitating all Europe. * * * These quotations will only whet the appetite of the scientific reader to devour the whole work. It is a book full of valuable information."—*Economist.*

"Decidedly a book to be read by all who take an interest in the progress of social science."—*Spectator.*

"A Southern man myself, never given to tariff doctrines, I confess to have been convinced by his reasoning, and, thank Heaven, have not now to learn the difference between dogged obstinacy and consistency. 'Ye gods, give us but light !' should be the motto of every inquirer after truth, but for far different and better purposes than that which prompted the exclamation."—*The late John S. Skinner.*

"A volume of extensive information, deep thought, high intelligence, and moreover of material utility."—*London Morning Advertiser.*

"Emanating from an active intellect, remarkable for distinct views and sincere convictions."—*Britannia.*

"'The Past, Present, and Future,' is a vast summary of progressive philosophy, wherein he demonstrates the benefit of political economy in the onward progress of mankind, which, ruled and directed by overwhelming influences of an exterior nature, advances little by little, until these exterior influences are rendered subservient in their turn, to increase as much as possible the extent of their wealth and riches."—*Dictionnaire Universel des Contemporains. Par G. Vapereau. Paris,* 1858.

Principles of Social Science.

Three volumes, 8vo., cloth......................................$7.50

CONTENTS.—Volume I. Of Science and its Methods—Of Man, the Subject of Social Science—Of Increase in the Numbers of Mankind—Of the Occupation of the Earth—Of Value—Of Wealth—Of the Formation of Society—Of Appropriation—Of Changes of Matter in Place—Of Mechanical and Chemical Changes in the Forms of Matter. Volume II. Of Vital Changes in the Form of Matter—Of the Instrument of Association. Volume III. Of Production and Consumption—Of Accumulation—Of Circulation—Of Distribution—Of Concentration and Centralization—Of Competition—Of Population—Of Food and Population—Of Colonization—Of the Malthusian Theory—Of Commerce—Of the Societary Organization—Of Social Science.

"I have no desire here to reproach Mr. Malthus with the extreme lightness of his scientific baggage. In his day, biology, animal and vegetable chemistry, the relations of the various portions of the human organism, etc. etc., had made but little progress, and it is to the general ignorance in reference to these questions that we must, as I think, look for explanation of the fact that he should, with so much confidence, in reference to so very grave a subject, have ventured to suggest a formula so arbitrary in its character, and one whose hollowness becomes now so clearly manifest. Mr. Carey's advantage over him, both as to facts and logic, is certainly due in great part to the progress that has since been made in all the sciences connected with life; but then, how admirably has he profited of them ! How entirely is he *au courant* of all these branches of knowledge which, whether

THE WORKS OF HENRY C. CAREY.

directly or indirectly, bear upon his subject! With what skill does he ask of each and every of them all that it can be made to furnish, whether of facts or arguments! With what elevated views, and what amplitude of means, does he go forward in his work! Above all, how thorough in his scientific caution! Accumulating inductions, and presenting for consideration facts the most undoubted and probabilities of the highest kind, he yet affirms nothing, contenting himself with showing that his opponent had no good reason for affirming the nature of the progression, nor the time of duplication, nor the generalization which takes the facts of an individual case and deduces from them a law for every race, every climate, every civilization, every condition, moral or physical, permanent and transient, healthy or unhealthy, of the various populations of the many countries of the world. Then, having reduced the theory to the level of a mere hypothesis, he crushes it to atoms under the weight of facts."— *M. De Fontenay in the "Journal des Economistes." Paris, September*, 1862.

"This book is so abundantly full of notices, facts, comparisons, calculations, and arguments, that too much would be lost by laying a part of it before the eye of the reader. The work is vast and severe in its conception and aim, and is far removed from the common run of the books on similar subjects."—*Il Mondo Letterario, Turin.*

"In political economy, America is represented by one of the strongest and most original writers of the age, Henry C. Carey, of Philadelphia. * * * * * * * *
"His theory of Rents is regarded as a complete demonstration that the popular views derived from Ricardo are erroneous; and on the subject of Protection, he is generally confessed to be the master-thinker of his country."—*Westminster Review.*

"Both in America and on the Continent, Mr. Henry Carey has acquired a great name as a political economist. * * * *
"His refutation of Malthus and Ricardo we consider most triumphant."—*London Critic.*

"Mr. Carey began his publication of Principles twenty years ago; he is certainly a mature and deliberate writer. More than this, he is readable: his pages swarm with illustrative facts and with American instances. * * * * * * * * *
"We are in great charity with books which, like Mr. Carey's, theorize with excessive boldness, when the author, as does Mr. Carey, possesses information and reasoning power."—*London Athenæum.*

"Those who would fight against the insatiate greed and unscrupulous misrepresentations of the Manchester school, which we have frequently exposed, without any of their organs having ever dared to make reply, will find in this and Mr. Carey's other works an immense store of arms and ammunition. * * * * * * *
"An author who has, among the political economists of Germany and France, numerous readers, is worth attentive perusal in England."—*London Statesman.*

"Of all the varied answers to the old cry of human nature, 'Who will show us any good?' none are more sententious than Mr. Carey's. He says to Kings, Presidents, and People, 'Keep the nation at work, and the greater the variety of employments the better.' He is seeking and elucidating the great radical laws of matter as regards man. He is at once the apostle and evangelist of temporal righteousness."—*National Intelligencer.*

"A work which we believe to be the greatest ever written by an American, and one which will in future ages be pointed out as the most successful effort of its time to form the great *scientia scientiarum.*"—*Philadelphia Evening Bulletin.*

PRACTICAL AND SCIENTIFIC BOOKS.

THE WORKS OF HENRY C. CAREY.

The Slave Trade, Domestic and Foreign;

Why it Exists, and How it may be Extinguished. 12mo., cloth...$1.25

CONTENTS.—The Wide Extent of Slavery—Of Slavery in the British Colonies—Of Slavery in the United States—Of Emancipation in the British Colonies—How Man passes from Poverty and Slavery toward Wealth and Freedom—How Wealth tends to Increase—How Labor acquires Value and Man becomes Free—How Man passes from Wealth and Freedom toward Poverty and Slavery—How Slavery grew, and How it is now maintained in the West Indies—How Slavery grew, and is maintained in the United States—How Slavery grows in Portugal and Turkey—How Slavery grows in India—How Slavery grows in Ireland and Scotland—How Slavery grows in England—How can Slavery be extinguished?—How Freedom grows in Northern Germany—How Freedom grows in Russia—How Freedom grows in Denmark—How Freedom grows in Spain and Belgium—Of the Duty of the People of the United States—Of the Duty of the People of England.

"As a philosophical writer, Mr. Carey is remarkable for the union of comprehensive generalizations with a copious induction of facts. His research of principles never leads him to the neglect of details; nor is his accumulation of instances ever at the expense of universal truth. He is, doubtless, intent on the investigation of laws, as the appropriate aim of science, but no passion for theory seduces him into the region of pure speculation. His mind is no less historical than philosophical, and had he not chosen the severer branch in which his studies have borne such excellent fruit, he would have attained an eminent rank among the historians from whom the literature of our country has received such signal illustration."—*New York Tribune.*

French Politico-Economic Controversy,

Between the Supporters of the Doctrines of CAREY and of those of RICARDO and MALTHUS. By MM. De Fontenay, Dupuit, Baudrillart, and others. Translated from the "Journal des Economistes," 1862-63. (*In press.*)

Protection of Home Labor and Home Productions

Necessary to the Prosperity of the American Farmer. By H. C. Baird. Paper..13

Smith. A Manual of Political Economy.

By E. Peshine Smith. 12mo., cloth......................$1.25

www.ingramcontent.com/pod-product-compliance
Lightning Source LLC
Chambersburg PA
CBHW020912230426
43666CB00008B/1419